Times of Terror

Times of Terror

Discourse, Temporality and the War on Terror

Lee Jarvis
Lecturer, Department of Politics and International Relations, Swansea University, UK

palgrave
macmillan

First published 2009 by
PALGRAVE MACMILLAN

Palgrave Macmillan in the UK is an imprint of Macmillan Publishers Limited, registered in England, company number 785998, of Houndmills, Basingstoke, Hampshire RG21 6XS.

Palgrave Macmillan in the US is a division of St Martin's Press LLC, 175 Fifth Avenue, New York, NY 10010.

Palgrave Macmillan is the global academic imprint of the above companies and has companies and representatives throughout the world.

Palgrave® and Macmillan® are registered trademarks in the United States, the United Kingdom, Europe and other countries

ISBN-13: 978-0-230-22369-1 hardback
ISBN-10: 0-230-22369-9 hardback

This book is printed on paper suitable for recycling and made from fully managed and sustained forest sources. Logging, pulping and manufacturing processes are expected to conform to the environmental regulations of the country of origin.

A catalogue record for this book is available from the British Library.

Library of Congress Cataloging-in-Publication Data

Jarvis, Lee, 1979–
Times of terror : discourse, temporality and the War on Terror/
Lee Jarvis.
p. cm.
Includes bibliographical references and index.
ISBN 978-0-230-22369-1 (alk. paper)
1. War on Terrorism, 2001- 2. Critical discourse analysis. 3. Time.
I. Title.
HV6431.J38 2009
363.325—dc22 2009013621

10 9 8 7 6 5 4 3 2 1
18 17 16 15 14 13 12 11 10 09

Printed and bound in Great Britain by
CPI Antony Rowe, Chippenham and Eastbourne

For my parents, with love and thanks

Contents

Preface

This book has been in the making for a long time, accompanying my own personal journey through the universities of Birmingham, Oxford Brookes, and Swansea. Although I have benefited immeasurably from the advice, insights, and knowledge of very many individuals at each of those institutions and beyond, I would like to extend a particular note of gratitude to the following for their help, support, and encouragement along the way: Stephen R. Bates, Claes Belfrage, David Hudson, Richard Jackson, Laura Jenkins, Donna Lee, Tim Legrand, Michael Lister, Emily Pia, Magnus Ryner, Laura J. Shepherd, Jill Steans, Daniela Tepe, Matthew Watson, and Paul D. Williams. I would also like to thank the ESRC for providing the initial funding that was required for my research with the award PTA: 030 2003 00614. And, at Palgrave Macmillan, I am most grateful indeed to Alexandra Webster and Renee Takken for their support and continuing assistance in seeing this project progress from proposal to finish. Finally, this book both draws on and extends the arguments presented in an earlier article "Times of terror: writing temporality into the War on Terror", *Critical Studies on Terrorism* 1(2), pp. 245–62, available through: http://www.informaworld.com. I am grateful to Taylor & Francis for their permitting my using this material.

On a more personal note, I would like to express my profound gratitude to those most important people without whose support this book would never have been completed. To Aya, whose generosity of spirit, incredible energy, and unwavering embrace of life has given me the determination to keep going with this project: thank you so very much, I couldn't have done it without you! To my wider family that have encouraged and supported me throughout – and particularly to John and Doreen for their reading of earlier drafts of this book – I am deeply grateful again. To my brother Greg and sister Kerri, for helping me stay focused on the important things in life: thank you both. And, finally, for sparking my interest in the political, and helping me sustain that interest in more ways than I remember, this text is my parents' as much as it is mine. If not complicit in its shortcomings or limits, they are equally responsible for any contribution it may make. For these reasons, this book is dedicated to them. I wish I had more to offer.

Lee Jarvis

1
Introduction

On 14 September 2001, President George W. Bush addressed a remembrance service at the American National Cathedral in Washington, DC. Called to commemorate the now-infamous attacks of three days prior, the service offered a highly visible performance of national togetherness (Silberstein, 2002: 43–55). With an array of prominent representatives from the American political and religious establishments all attending the event, the sense that something remarkable – something exceptional – was being remembered seemed almost inescapable.

When his own turn to speak duly arrived, the President opened his address with a relatively predictable yet powerful offering of sympathy and prayer for the victims and their families that had been most directly affected by those attacks. Turning, then, to the rather more difficult task of reflecting on the attacks' significance and his administration's likely response, Bush proffered the following remarks:

> Just three days removed from these events, Americans do not yet have the distance of history. But our responsibility to history is already clear: to answer these attacks and rid the world of evil. War has been waged against us by stealth and deceit and murder. This nation is peaceful, but fierce when stirred to anger. This conflict was begun on the timing and terms of others. It will end in a way, and at an hour, of our choosing.
>
> (Bush, 14/9/2001a)

The President's comments are arresting for the sense of absolute moral and political conviction that permeates through this short passage. That they were delivered in such proximity to the attacks of

11 September 2001, indeed, only adds to the impression of unfolding drama that he captures so powerfully here. Yet for those of us who watched his administration both declare and institute 'War on Terror' in response to those events, the tone and content of these remarks will also be instantly, depressingly, familiar. As we now know all too well, impassioned discussions of national unity and innocence, emotive demands to eradicate evil, and a continued reaffirmation of American politico-military supremacy have rarely travelled far from governmental discussions of this peculiar conflict. And, with that thought in mind, these remarks are clearly of interest as an early example of the aggressive, self-confident rhetoric that has now long characterised the Bush administration's engagement with political violence. If not, necessarily, paradigmatic of the War on Terror's structuration, they are certainly, and vividly, indicative of that conflict's manufacture and sale from its earliest unfolding and some distance beyond.

For the purposes of this book more specifically, I think there is another, equally compelling, reason for taking these comments quite seriously. For, if we reread Bush's remarks with patience and care, we very quickly encounter a similarly central, if rather more camouflaged, theme that, again, rarely moved far from the unfolding War on Terror. By invoking history as a privileged medium for reflection and judgement: 'the distance of history'; by constructing history as a subject to which obligations are owed: 'our responsibility to history'; and, finally, by speaking with such sureness, such certainty, of events yet to come: 'it will end in a way, and at an hour, of our choosing', Bush's account is here thoroughly saturated with references to the historical and temporal. His account is saturated, more precisely, with a whole series of claims to a privileged knowledge of the dynamics and movements of time and history. And, as we shall see in the chapters that follow, these remarks were far from exceptional in this sense.

My aim in the discussion that follows is to thrust forward these practices and representations of temporality for analysis and critical reflection. In so doing, I hope to demonstrate that specific writings of time – specific writings of pasts, presents, and futures – were absolutely central to the perceived coherence, necessity, and legitimacy of the Bush administration's new War on Terror. Exploring this centrality, I will argue, tells us something quite interesting about the frequently complex relationships that link temporality, politics, discourse, and violence. Not only in these particular times of terror, but also beyond.

The (Long) (Global) War on Terror(ism)

Embarking on any form of critical encounter with the George W. Bush administration's War on Terror is a rather less straightforward task than it may at first appear. In the first instance, any engagement of this sort immediately stumbles across the difficulty of knowing how to name this rather peculiar entity. If Freedman (2002: 61) is correct in noting that wars are typically signified retrospectively, as surely he must be, this particular conflict has already been denoted or named in a number of ways by its creators and protagonists.

In the beginning, this was a war, simply, on or against terrorism. Despite the difficulties inherent to a declaration of war on an abstract noun of this sort (Glover, 2002: 208; Ivie, 2005: 140; Roberts, 2005: 124–5), this initial nomenclature at least benefited from some measure of rhetorical consistency. Although this consistency so frequently failed to migrate to the objects on which the terrorism label was conferred, we seemed somehow to know where we stood in both discussing and contesting this conflict.

Not long into this war, however, onlookers and critics witnessed an important discursive shift in its framing. That shift, of course, related to the linguistic contraction of terrorism to terror: a contraction that permitted a counter-intuitive, yet useful, expansion of this conflict's particular targets. With 'War on Terror' thus declared, not only were 'we' (whoever 'we' were) confronting an enemy of terrorists, their facilitators, and abetters, but rogue states, transnational criminals, and other, similarly terrifying, purveyors of fear had also suddenly entered the fray. As Paul Williams (2008: 10–11) has neatly suggested, this conflict has acted as something of a black hole capable of pulling ever more issues and challenges into its reach.

More recently still, the even more impressively ambiguous 'Long War' seems to have become the favoured term for discussing this entity, at least among American political elites. Named in this way, Bush's administration appears to have finally resisted the temptation to delimit the present and future targets of this conflict: falling back, simply, on an appeal to its likely longevity. While this more recent invocation of temporality returns us quite nicely to Bush's National Cathedral address with which we began, the remainder of this book will employ the War on Terror moniker in discussing this conflict. In the first instance, this preference offers our study both consistency and simplicity in usage. It is, of course, a rather less awkward presentation than that offered in the subheading used at the top of this section. More important, however, is this name's status as that seemingly preferred by its founders throughout the first three years of its existence. Where it is this period specifically that interests me in the discussion that follows, my employment of this term offers an attempt at fidelity to this framing: an attempt at fidelity that is crucial, I will argue, if we are to take this conflict seriously.

If the assortment of names conferred on this war throws up immediate logistical difficulties for our discussion, a further complexity emerges from the similarly dynamic character of that which here is our focus. As outlined in the chapters that follow, the War on Terror was already being imagined, described, and enacted as a 'war on many fronts' from the very earliest days of its inception: a war emphatically not reducible to its military moments however much attention they garnered for their violence and destructiveness. Although, no doubt, central to the administration's (equally fluid and dynamic) politico-strategic objectives, the high-profile, controversial, operations in Afghanistan, Iraq, and beyond do not, clearly, exhaust the parameters of this conflict. In this sense, to do (analytical) justice to the War on Terror as an object of critique, it is necessary that we engage with the veritable panoply of techniques, practices, and decisions that have structured it both at home and abroad. Although it is difficult, we must reflect as fully as possible on the legislative, diplomatic, militaristic, and other components of this conflict, tracing the relations between the USA PATRIOT Act, Guantánamo Bay, and Operation Enduring Freedom to name three prominent examples. Indeed, as we shall see later, it was frequently through specific (and, importantly, contestable) writings of temporality that these otherwise disassociable moments were able to be linked or cohered into one seemingly coherent political, strategic, discursive totality: one political, strategic, discursive totality that became, simply, the War on Terror.

If remaining attentive to this discursive and political fluidity renders conceptualising and critiquing the War on Terror a difficult task, neither of these dynamics should discourage our engagement. As we know from a host of contemporary theorists, the social and political worlds into which we are all thrown necessarily exist as messy, disorganised, and complex affairs. However desirable order, fixity, or closure may appear to students of the social, the worlds that we encounter typically resist these ideals. From such a perspective, a substantial element of our responsibility as analysts of these worlds is surely to engage with this mess: to locate and explore moments of movement and flux alongside those attempts to contain, simplify, or tame those dynamics. Although it is difficult, we must resist the temptation of a static, synchronic, mode of analysis in bringing these issues forward for reflection. And, from such a perspective, the sense of political and rhetorical dynamism that has accompanied the War on Terror alone makes this conflict already an interesting problematic for study.

A second, more urgent, reason for engaging with this war in spite of the difficulties noted above relates to the well-documented, and pernicious, implications that have followed in the War on Terror's wake. With a little over seven years having passed since its commencement as I write this passage, its incredible human cost shows very few signs of meaningful abatement. With one prominent source reporting a total of over 80,000 civilian deaths in Iraq alone by June 2008 (IBC, 2008), we have clearly witnessed the launch of a particularly bloody and violent period of global politics. Although there are considerable difficulties – analytical, and not least political – involved in determining the precise numbers of casualties in any context (Lacina & Gleditsch, 2005: 146), engaging with this violence becomes even more urgent once we take note of the deaths in Afghanistan and beyond.

Although a less immediately bloody consideration, it is important, also, not to underestimate the more pervasive costs of this war on established and long-held international norms: however imperfect those norms may indeed be. As critics have noted, the emergence, indeed celebration, of preventative warfare as the cornerstone of the Bush doctrine offers an unfortunate example of the ease with which expedience may trump normative concerns when the politics of emergency and exceptionality is invoked or even trumpeted (see Chomsky, 2003: 13–16; Bellamy, 2008: 107). And, domestically, of course, we have witnessed a whole raft of equally deplorable incursions on embedded civil liberties within the United States and beyond. With the suspension of habeas corpus and other rights for those now designated 'enemy combatants' (Butler, 2006: 50–100; Isin & Rygiel, 2007: 197; Weber, 2007: 115), the emergence of extraordinary renditions as a security strategy (Mutimer, 2007), and a renewed quest for ever more invasive techniques of surveillance (Lyon, 2003: 13; Smith, 2008: 175), the War on Terror has run roughshod over established civil, political, and legal practices alike. There is perhaps no greater indictment of this conflict, indeed, than the vigour it has given those keen to discuss liberal democratic torture as a legitimate counterterrorist policy (see, for example, Žižek, 2002: 102–6; Bellamy, 2006; Kennedy-Pipe & Rengger, 2006: 550; Lukes, 2006). And, if we are to take these developments seriously, as again surely we must, we simply cannot eschew engagement with this newest War on Terror in all of its complexity.

With these preliminary remarks behind us, I should like to continue our discussion with a brief sketch of my own attempt to understand the Bush administration's War on Terror in its fluidity. Where my purpose in this book is to engage with the representations and performances of temporality that have thoroughly saturated this conflict, it will be useful to begin by clarifying how I conceptualise this entity here. With this achieved, we shall then be in a better position to embark on our exploration of the relationship between discourse, politics, temporality, and violence introduced briefly above.

The War on Terror as political discourse

In contemporary social scientific literatures we have witnessed a considerable surge of interest in the concept of discourse. This book both draws on and contributes to these discussions by offering this concept as a particularly valuable lens through which to read the Bush administration's War on Terror. Because the contemporary popularity of this term is matched by its contestability (see, for example, Purvis & Hunt, 1993; Howarth, 2000; Mills, 2004), it will be useful to briefly clarify the specific understanding employed in the investigation that follows: not least to pre-empt several well-known, and well-worn, objections to it. To these ends, let us now turn to Ernesto Laclau's approach to discourse as a productive broad framework from which to explore the significance that claims to temporality have performed within this particular politico-strategic problematic.

In common with many contemporary theorists, Laclau formulates his understanding of discourse against the assumption that our social world does not possess any foundation or definitive ground (Laclau & Mouffe, 2001: 95–96; Laclau, 2005: 68–9; Critchley & Marchart, 2004: 4). Put simply, this assumption suggests that all of the taken-for-granted components of our existence(s) – all of the identities, institutions, social practices, and so forth that structure our lives – have come into being only as the outcome of a complex process of historical play, contestation, and struggle (Torfing, 1999: 62). However strongly we may desire it, there simply does not exist any ultimate logic to which we can turn to explain how the world has been structured as it appears to us. The usual suspects so frequently called on to fulfil this explanatory role – Reason, History, God, Man, and so forth – themselves are nothing more than human constructions. There is no universal story, no 'metanarrative' (Lyotard, 1984: xxiv), capable of linking our pasts, presents, and futures into the coherent totalities for which we may intuitively search. All we have is a patchwork of precarious and ultimately contingent – of undetermined and ultimately unpredictable – patterns of life. A patchwork of social frameworks subject to challenge and change however stable they may appear to those caught up within them. There is, in short, nothing inevitable, natural, or necessary underpinning the ways in which our social worlds are ordered and lived.

This theorisation of a foundationless existence poses an exciting and potentially quite serious challenge to any number of established political perspectives based on an illusion of metaphysical depth. In order to understand the processes through which our social horizons come into being, however, something more is clearly needed here. To understand the ways in which those horizons also constrain, encourage, and enable particular behaviours, again more needs to be said. None of us inhabits a world of perpetual instability: there are always more or less established – more or less accepted – rules, norms, and practices that structure the social contexts in which we are located. My world, and my place in that world, will always appear more or less sedimented, more or less fixed, however much I may wish it were otherwise. What is needed, therefore, is a way of understanding these contexts: a way of conceptualising those moments of fixity *and* transformation in a foundationless world. For, as Laclau (1990: 27) suggests, 'It is true that in the *final* instance no objectivity can be referred back to an absolute ground; but no important conclusion can be drawn from this, since the social agents never act in that final instance.'

Laclau's response to this gap in our understanding of the construction of specific social existences centres on his particular reading of the concept of discourse. In his own, broadly post-Marxist, approach, discourses or discursive structures refer to the partial and temporary moments of stability that emerge when particular social agents succeed in their attempts to stabilise our undecidable ontological horizon (see, for example, Howarth, 2000: 101–2; Laclau & Mouffe, 2001: 105; Hajer, 2005: 300–1). Approached in this way, discourses are viewed as relatively coherent and meaningful structures or systems that both constitute and organise social agents, institutions, and objects around their perceived similarities and differences (Laclau & Mouffe, 2001: 105). In Hajer's (2005: 300) useful summary, 'Discourse is defined here as an ensemble of ideas, concepts, and categories through which meaning is given to social and physical phenomena, and which is produced and reproduced through an identifiable set of practices.' As noted above, I approach the War on Terror as one such discourse – one such meaningful system of relations – throughout the remainder of this book.

Thinking about the War on Terror in this way is useful, I suggest, for three reasons that will be of particular interest to our study. In the first instance, this Laclauian conception of discourse refuses to posit any ontological distinction between the material and non-material moments of a particular social context (Laclau, 1989: 71; Laclau, 2005: 78; Laclau & Mouffe, 2001: 108). Social subjects and objects – *and* our ideas about those subjects and objects – are viewed from this perspective as inherently, intrinsically, related. We need not, we cannot, make any arbitrary a priori decision to focus on the one at the expense of the other. This is significant for our purposes, in that it enables our tracing the relations that exist between the War on Terror's linguistic and non-linguistic dimensions. It enables us to explore not only the ways in which the PATRIOT Act, Guantanamo Bay, and Operation Enduring Freedom, for example, were justified within official rhetoric but also the ways in which they were constituted and understood as meaningful components of the same discursive framework. And yet, as Laclau and Mouffe (2001: 108) argue in a much-cited passage, viewing the social through this discursive lens is in no way to reduce our analysis to the study of language and ideas alone. Nor is it to deny the existence of material 'reality', for:

> [t]he fact that every object is constituted as an object of discourse has nothing to do with whether there is a world external to thought, or with the realism/idealism opposition. An earthquake or the falling of a brick is an event that certainly exists, in the sense that it occurs here and now, independently of my will. But whether their specificity as objects is constructed in terms of 'natural phenomena' or 'expressions of the wrath of God', depends upon the structuring of a discursive field. What is denied is not that such objects exist externally to thought, but the rather different assertion that they could constitute themselves as objects outside any discursive conditions of emergence.

In short, Laclau's approach to the study of discourse offers a useful conceptual apparatus through which to explore the interaction of ideas, material objects, and agents at the particular historical juncture that has come to be known as the War on Terror. It will enable our taking seriously the ways in which particular events such as 9/11 were interpreted and reproduced within specific conceptions of time and history, and to trace the implications of those interpretations for the whole panoply of responses that were subsequently instituted under this discourse's rubric. For, as I hope to demonstrate in the discussion that follows, it is indeed perfectly possible to take this language of time and history seriously without in any way denying a material reality 'beneath' or 'below' it.

A second reason for conceptualising the War on Terror in this way relates to Laclau's argument that any discursive structure or formation will always, necessarily, remain incomplete. If our socio-political worlds of institutions, identities, and so forth are constructed upon a radically undecidable horizon without foundation or ground, it follows that no attempt to capture or stabilise that horizon will ever be fully and finally successful. However stable or coherent any systemic organisation of subjects and objects appears at a particular historical juncture, its rules, norms, practices, behaviours, and so forth will all have the capacity to be reperformed or reinterpreted differently. They will, put otherwise, all have the capacity to be transformed or changed over time despite their longevity or apparently natural character (see, for example, Laclau, 1990: 29; Torfing, 1999: 62–6). This recognition of structural incompleteness is important for our engagement with the War on Terror as it allows us to take seriously those moments of porosity, heterogeneity, and apparent contradiction that have characterised this relatively bounded totality from its inception. Although, perhaps, easier to approach this formation as a coherent and unified structure, exploring these moments offers a far more useful perspective from which to highlight and critique this discourse's contingent – and therefore contestable – existence. To offer a brief example, rather than attempting to adjudicate between governmental writings of 9/11 as unprecedented attack and Pearl Harbor revisited, simply drawing attention to these heterogeneities offers a productive and powerful strategy for critiquing the sense of inevitability and necessity that has long accompanied the War on Terror.

The final advantage of employing Laclau's broad framework for conceptualising the War on Terror concerns the explicitly political emphasis underpinning his conception of discourse. Where our daily lives are always already structured by a series of sedimented or familiar behaviours, practices, and expectations, Laclau approaches these taken-for-granted realities as the realm of the social. The social, understood thus, refers to the world(s) of discourse into which we are all thrown as individuals: a world of constraints and enabling conditions against which all our actions occur. The political, in contrast, refers to those operations of power, those creative acts or decisions, through which social and discursive contexts are themselves reconstituted, reinstituted, or reproduced (Laclau, 1996: 47). It is the political, in short, that ensures our social world 'is not entirely defined in terms of repetitive, sedimented practices' (Laclau, 1994: 3).

Thinking about the political in terms of those radically constructive decisions or moments that underpin historical emergence and change again poses considerable relevance for our engagement with the War on Terror. By approaching the formulation and formation of this war as an inherently political dynamic – neither necessary nor predetermined – we will once again be able to take seriously – to critique – the claims to necessity around which it has been constructed and justified. By approaching it, put otherwise, as the product of political contestation and decision, we will be able to render the necessary visible as contingent (Watson & Hay, 2003): to draw attention to the centrality of power in the creation of this seemingly coherent and inevitable entity. With the War on Terror viewed thus as an effort (material, incomplete and far from inevitable) to stabilise one particular dimension of our social world, we will be able to also explore the exclusion of a whole range of alternative and equally valid responses to 9/11 that this particular discourse militated so forcefully against. For, as Torfing (2005: 23) suggests in a recent discussion:

> discourse theory puts power and power struggles at the top of the agenda. Power is not analysed in terms of a resource or capacity one can possess, store, or retrieve, or as a relation of domination. Power is conceived in terms of the political acts of inclusion and exclusion that shape social meanings and identities and condition the construction of social antagonisms and political frontiers. The construction of discourse always involves both inclusion and exclusion of identity and this means that discourse and power are intrinsically linked with each other.

By exploring the centrality of temporal claims within the War on Terror, then, this book offers one attempt to take the inherently contingent and political character of this conflict seriously. It is an attempt, put otherwise, to draw attention not only to moments of internal heterogeneity and tension within this particular war but also to excluded possibilities of understanding and action from without.

To summarise briefly, the following study approaches the George W. Bush administration's War on Terror as a meaningful and ostensibly coherent social construction: a political discourse in the Laclauian sense. By exploring the significance of particular, and distinct, temporal claims in the production and legitimisation of this conflict, I hope, first, to demonstrate that this totality of material and non-material moments was neither as complete nor as natural as it so frequently appeared. And, as outlined later, I hope also to demonstrate the discursive productivity and value that claims to temporality may, indeed do, perform in a far broader sense.

Time for critical terrorism studies

My own interest in the War on Terror taps into a well-documented and growing academic concern with the politics of terror more broadly (see, for example, Kushner, 2002: xii; Turk, 2004: 271; Gunning, 2007: 363; Jarvis, 2009). Although traditionally seen as something of a minority pursuit within more established disciplinary frameworks (Merari, 1991; Gordon, 1999; Richardson, 2006: 1), the threat of unconventional violence today commands considerable scholarly attention. With dedicated conferences, journal special issues, and targeted book series all increasingly familiar indicators of this burgeoning interest, the global impact of terrorism today offers a remarkable growth industry within social scientific research. As Gunning (2007: 363) has noted, 'On entering almost any bookshop, one is overwhelmed by the number of books discussing the war on terror, "Islamic terrorism" or "terrorism" more generally. Conference papers on "terrorism" abound and interest in issues related to "terrorism" has increased dramatically among scholars in cognate disciplines; degree programmes have been set up; funding opportunities have increased.'

For the purposes of this book, one of the more interesting developments within this expanding field of study has been the emergence of a literature explicitly concerned with representations, narratives, and performances of (counter)terrorism. Building on earlier works exploring the relationships between discourse, identity, and political violence (see, for example, Campbell, 1998a, 1998b; Neumann, 1999; Said, 2003), these recent discussions have been particularly productive in tracing the ways in which terrorism is produced as an object of knowledge, an identity, and a threat at specific historical junctures. Despite significant differences of epistemological and normative emphasis, these discrete analytical projects cohere around a common attempt to reveal, contest, and ultimately disrupt specific framings of terror. Examples to date of this 'interpretivist face of critical terrorism studies' (Jarvis, 2009) have included, on the one hand, a number of broad overviews of the War on Terror's construction across distinct discursive sites (see, for example, Collins & Glover, 2002; Silberstein, 2002; Jackson, 2005; Lewis, 2005;

Croft, 2006; Jarvis, 2008). And, on the other hand, more localised discussions of the power that specific tropes and narratives of terror possess in contemporary political and popular discourse (see, for example, Morris, 2004; Devetak, 2005; Jackson, 2007b).

Despite already attracting both hostile and sympathetic critique (on this, see Jackson, 2007a; Horgan & Boyle, 2008), this broad body of literature offers a thought-provoking challenge to the more traditional concerns of those working within the field of terrorism studies. If unconventional violence has indeed historically suffered from scholarly neglect, the research that had taken place in this area had, until rather recently, been remarkably narrow in emphasis. Specifically, and with surprisingly few exceptions (see, for example, Feldman, 1991; Zulaika & Douglass, 1996), it has been animated by three intimately related and long-standing questions: how to define terrorism, how to explain terrorism, and how to prevent it (Jarvis, 2009). As such, by exploring not terrorism per se, but, rather, *productions* or *constructions* of terrorism, these literatures open significant space for rethinking not only our object of study – (counter)terrorism – but also our own role as students of terror.

In the first instance, by tracing the historical, social, and discursive processes through which events and actors become inserted into the rubric of terrorism, this interpretivist research has been vital in demonstrating the contingent character of these constructions. However disturbing or discomforting it may appear, it is crucial we remember that discussions of trauma, evil, cowardice, and the like are borne not of necessity or ontological grounding, but rather, and emphatically, of the choices of particular actors. Whether these choices are deliberate or otherwise – whether they are strategic decisions or ill-considered reactions to violences – alternative readings and understandings of any event are always, necessarily, possible. Through bringing the contingency of existing narratives of terror sharply into focus, then, these literatures offer a powerful route into questioning the claims to objectivity and truth that so frequently accompany discussions of political violence.

By drawing our attention to the contingency underpinning accepted – seemingly objective – discourses of terror, these literatures also perform a second function of considerable importance. Following, in part, Foucault's earlier work on problematisation (1984: 388–90), they open space for challenging the congruity or fit between formulations of problems and responses within the context of political violence. As Jackson (2005: 38–40) and others (Silberstein, 2002: 14; Dittmer, 2005: 638) have argued, alternative framings of 9/11 as an act of criminality rather than war, for example, would likely have engendered vastly different policy solutions to the militaristic, state-centric emphasis that characterised the subsequent War on Terror. And, by demonstrating the legitimatory functions of that particular framing and others, these discussions enable our contesting the frequently pernicious and damaging linkages that exist between terror and counterterrorist violences. Not from a privileged Archimedean fulcrum based on a claim to superior knowledge. But, rather, through a recognition of the unstable and contestable character on which this mimetic cycle of violence depends.

By facilitating a critical exploration into specific framings of (counter)terror and their consequences, these literatures offer one final insight of value for our discussion. Where terrorism was traditionally viewed as an objective problem for scholars to define, explain, and ultimately resolve, that particular paradigm mobilised a remarkably restrictive conception of academic responsibility (Jarvis, 2009). Approached in this way, the role of the student of terror was always already delimited in advance to one of risk identification and management. Our function, simply, to assist policymakers in their efforts to formulate ever more efficient responses to this threat. In this sense, as noted by Gunning (2007), the study of terrorism has typically functioned as a problem-solving pursuit in the Coxian formulation, a problem-solving pursuit that:

> takes the world as it finds it, with the prevailing social and power relationships and the institutions into which they are organized, as the given framework for action. The general aim of problem solving is to make these relationships and institutions work smoothly by dealing effectively with particular sources of trouble.
>
> (Cox, 1996: 88)

What these discursively oriented literatures offer, in contrast, is a commitment to scholarship that is wedded not to an ameliorative, problem-solving defence of the status quo, but a commitment to scholarship wedded to critical enquiry, of 'saying the unsayable' in Booth's (2008: 68) terminology. As is, hopefully, already clear, this book offers an attempt to continue and extend this critical emphasis.

Times of terror

The remainder of this book contributes to this growing interest in the relationship between discourse, terrorism, and counterterrorism. As suggested above, it does so by exploring the emergence and import of particular conceptions of temporality within the Bush administration's unfolding War on Terror. Claims to a privileged knowledge of the movement, dynamics, and patterns of time, I will argue, were absolutely fundamental to the creation, representation, and selling of this conflict in at least four related ways.

In the first instance, I will argue that specific representations of time offered a powerful and productive discursive resource for inscribing *significance* not only into key developments within the ongoing War on Terror but also into this conflict as a totality. Whether this significance emerged out of claims to absolute historical uniqueness – absolute singularity – as was frequently the case in discussions of 9/11. Or, indeed, whether through claims to a discernible dynamic of linear progression – as in discussions concerning advances made in the fight against terrorism. Specific conceptions of temporal movement and change were frequently central to the Bush administration's efforts at writing (political, strategic, and normative) urgency into this conflict.

Second, I will argue that specific constructions of temporality were also central to the inscription of political and moral *legitimacy* into the War on Terror as a whole and its constituent parts. On occasions, this legitimacy emerged out of governmental appeals for unprecedented mechanisms to counter the new, exceptional, threat being faced by America and its allies. On others, from appeals to the lessons learned within earlier (unsuccessful) efforts at countering global terrorism. And, in other cases still, from claims to the existence of a timeless collection of moral precepts and rules that not only prohibited certain (terrorist) behaviours but also legitimised certain (counterterrorist) responses. Whatever the example, to fully understand the persuasiveness of this particular conflict to American and global audiences, it is necessary once again to engage with its constitution across particular temporal imaginaries.

In the third instance, I hope also to demonstrate that specific representations of time offered an important discursive mechanism for introducing *structural coherence* into the War on Terror as a meaningful totality. By cohering or linking discrete and otherwise disassociable political developments around shared temporal dynamics (of evolution, revolution, *and* timelessness), writings of time were central in concealing the above-noted structural incompletion of this political discourse. As we shall see, that this conflict appeared as a coherent, unified, and consistent response to the attacks of 11 September 2001 depended, in large part, on its construction around specific claims to a privileged knowledge of temporality.

Finally, I will also argue that specific writings of temporality were vital in reproducing many of the collective identities that emerged as integral to the construction of this unfolding war. At the level of the self, appeals to a common past, present, or future were crucial in interpellating American citizens into a coherent social grouping with a shared, tangible, stake in the outcome of this conflict. Similar appeals to a shared (hemispheric, civilisational, or even global) past, present, or future, indeed, also offered a key strategy for extending this war's parameters and import far beyond the United States' borders. At the level of the other, efforts to parallel *and* distinguish al-Qaeda from other past, present, and future dangers were similarly critical in concretising this particular threat and its equivalents. Whether viewed as radical departure from earlier terrorisms or as extension or repetition of other previous threats, the Bush administration's persistent appeals to imminent, continuing, danger were thoroughly saturated once more with appeals to the temporal. To fully understand the *identity politics* of the War on Terror, then, we again need to take seriously its writing around specific conceptions of time and history.

In sum, this book offers an attempt to trace the ways in which particular claims to a privileged knowledge of temporality were central to the perceived necessity, legitimacy, coherence, and identity politics of the War on Terror. By exploring these dynamics, I hope to offer four contributions in the discussion that follows.

In the first instance, despite the advances of the above-noted interpretive literatures, much of the critical work conducted in this field has been rather slow in engaging with the constitutive role of temporality in discourses of terror. Although existing accounts of the War on Terror have successfully highlighted the widespread positioning of 9/11 as a moment of temporal rupture (Campbell, 2002; Cox, 2002: 261; Der Derian, 2002: 104; Derrida, 2003: 85–90; Jackson, 2005: 29–38; Bousquet, 2006; Kennedy-Pipe & Rengger, 2006: 539), and the tendency towards selective historical analogising within the Bush administration's representations of this war (Noon, 2004; Jackson, 2005: 41–4; Croft, 2006: 55), there has yet to emerge any sustained engagement with the far more pervasive functions performed by reproductions of time in this context. As such, this book offers an attempt to open a further, and critical, dimension of this particular conflict up for reflection, scrutiny, and analysis. Bringing to light hitherto-neglected issues and concerns through a sustained, patient reading of this sort, I suggest, offers an important starting point for any attempt at engaged social analysis (see Foucault, 1984; Ricoeur, 1999: 16; Jarvis, 2008: 245–6).

The second contribution of this book builds directly on the first. As I have already hinted, if the War on Terror was constructed around, and legitimised by, claims to temporality from its inception, it was also constructed around, and legitimised by, radically *distinct* claims to temporality. As we shall see in the following chapters, the governmental representatives responsible for this conflict oscillated between very different understandings of time's movements and shapes in discussing and justifying this ongoing war. And, by drawing attention to the ways in which events and identities such as 9/11, Americans, and al-Qaeda were inserted into different, indeed incompatible, constructions of time, this book offers an alternative, and as yet underexplored, space from which to critique this war's apparent coherence. In this sense, rather than attempting to assess the accuracy of the Bush administration's claims to temporal discontinuity, linearity, and timelessness, I will argue that simply drawing attention to the concurrent employment of these imaginaries offers a prescient reminder of the War on Terror's ultimate lack of fixity or metaphysical presence. This internally oriented effort at destabilising the War on Terror, of course, requires our thinking rather carefully about the status and practice of critique. I return to this theme later.

The third contribution of this book poses arguably greater significance than the two mentioned thus far. While exploring the War on Terror's temporal framing tells us a great deal about that particular conflict, it tells us rather more about the politics of identity, security, and violence more broadly. As we shall see in the chapters that follow, perceptions of who we are, what we are – be that 'Americans', 'civilised', 'freedom-loving', or whatever – are necessarily, inherently, rooted in specific understandings of temporality. Perceptions of whom and what we fear – be that 'new terrorism', 'evildoers', 'barbarians', and so forth – emerge, similarly, from particular engagements and writings of time. By tracing the significance of temporal claims to the emergence of specific individual and collective subjectivities, then, this book offers an attempt to rethink our understanding of the practices through which identity, security, and violence are enacted and performed in contemporary political life.

The final claim I would like to make for the analysis that follows is, perhaps, the most ambitious of all. For, as I shall argue in the course of our discussion, exploring the writing of the War on Terror around claims to temporality also sheds considerable light on the relationship between temporality and the political. As indicated already, I associate the political here not only with those moments of radical institution wherein our social worlds come into being through struggle and contest (see, for example, Laclau, 1996: 47; Edkins, 1999: 2, 2003: 13–14) but also with those processes of exclusion, antagonism, and othering that *necessarily* accompany attempts to institute the social in particular ways (see, for example, White, 1990: 62–3; Mouffe, 2005: 10–14). With this in mind, I will argue that time itself is always inherently, necessarily, political in each of these senses. First, by demonstrating the centrality of temporal claims within the reconstitution of the War on Terror, I will suggest that invocations of time always perform a radically constructive role: a radically constructive role that serves to enable or bring into being all of the identities, institutions, and so forth that comprise our social existence. In the second, I will also demonstrate that invocations of time – any invocation of time – work through the exclusion of alternative, equally plausible, understandings of temporality – and, therefore, subsequently, equally plausible responses to the passage of time. In short, I will argue that it is simply impossible to conceive of temporality and the movement of history shorn of any notion of power and the political.

To summarise briefly, my argument in this book is that particular representations of temporality were central to the Bush administration's efforts to render the War on Terror possible, legitimate, coherent, and meaningful. Tracing these representations and their outcomes, I will suggest, not only adds further nuance to the burgeoning interest in the relationships between discourse and (counter)terror that we have encountered but also offers an underexplored perspective from which to critique this particular formation, and, subsequently, to rethink the character of identity, security, violence, temporality, and the political more broadly. Before embarking on this exploration, however, I would like now to conclude this chapter by reflecting very briefly, first, on some matters of methodology and, second, on the structure of the remainder of the book.

Methodological reflections

One of the most immediate obstacles confronting those seeking to understand the Bush administration's War on Terror is the sheer volume of potentially relevant material available. The digital age in which we now live affords texts (speeches, policy documents, photographs, films, and so forth) a life and accessibility that extends far beyond the initial conditions and contexts of their production. While undoubtedly exciting for any researcher seeking to open up archives and force subjugated knowledges into the foreground, this spatio-temporal stretching of texts clearly introduces new methodological problems and responsibilities concerning our selection and management of material. With this in mind, it will be valuable, briefly, to sketch the institutional and temporal parameters delimiting the research on which my arguments are based.

My own account of the significance and functions of temporality within the War on Terror political discourse is drawn from an analysis of over 600 texts produced by key representatives of the Bush administration in relation to this unfolding conflict. These – primarily linguistic – texts include speeches, testimonies, interviews, policy documents, and commentary pieces emanating from five sites of government: The White House, Department of State, Department of Defense, Department of Justice, and, Office/Department of Homeland Security. And, although my arguments trace the chronological continuation of specific scriptings of time up to 2007, the focus here is restricted to particular claims to the temporal – particular claims to pasts, presents, futures, and their relations – that emerged within the first three years of the War on Terror alone.

My focus on these five governmental departments is justified on two grounds. In the first instance, the leading figures of these departments were undoubtedly the most prominent in constructing and selling this war. Not only did these individuals demonstrate a persistent and repeated inclination to refer to the unfolding War on Terror in media opportunities, policy documents, and the like. They were also particularly vocal and visible in espousing their own significance concerning its conduct and likely future successes. And, by incorporating figures from the executive, diplomatic, military, and legislative elements of this 'war on many fronts', I have attempted to offer as broad an understanding of War on Terror as possible. Focusing on the output of the White House or Department of Defense alone, for example, would clearly leave significant dimensions of this war underexplored. In the second instance, my focus here on the quite voluminous output of these five sites of government affords our discussion quite sufficient material for a comprehensive understanding of the relationship between temporality, identity, security, violence, and the political, both in this specific context and beyond.

My focus on narratives of temporality that emerged within the first three years of the War on Terror alone represents a similar negotiation of empirical breadth and analytical depth. The most relevant factor for my restricting the material engaged with in this way relates to the remarkably productive and dynamic character of the Bush administration between the 9/11 attacks and the end of 2004. As we already know, that specific time period not only witnessed the launch of considerable and deeply controversial military operations in Afghanistan and Iraq but also saw the construction of a new Office and then Department of Homeland Security, and the passage of the 2001 USA PATRIOT Act as key elements in the domestic fight against terrorism. Furthermore, as we shall see in the chapters that follow, this period also witnessed a particularly productive moment in that administration's imagining, writing, and calling on temporality. With three entirely distinct conceptions of time resounding forcefully and continually throughout those three years, this focus again offers a vast quantity of relevant material with which to better understand the politics of time within this particular historical problematic.

Studying discourse: Discover, recover, invent!

With the textual and chronological parameters of the book established, it will be useful to conclude our methodological reflections with a brief note on matters of epistemology. This is important, I suspect, for, even if we can agree that language or discourse perform a significant role in the politics of terror, temporality, and so forth, clarifying the specific status and functions of that language requires still further reflection. What, specifically, are we to look for in our reading of the Bush administration's claims to a knowledge of time? Through what lens are we to assess or analyse the texts drawn on in the discussion that follows?

At the risk of considerable simplification, studies of discourse and language may be usefully divided into three interpretive analytics or styles of enquiry. The first of these, which I shall refer to here as *discursive discovery*, relates to those efforts at locating the truth of specific discursive practices – the truth of particular statements, performances, and so forth – as potentially independent from the relevant social actors' self-understanding of their practices. Approached in this way, the task of those studying discourse is to reveal, chart, and explain sites of incongruity between the world of language and the world as it is: to discover and revel in hypocrisy, misinformation, and the other inconsistencies that so frequently characterise the output of those speaking truths of the world. Paul Ricoeur's description of a 'hermeneutics of suspicion' captures the voyage of discovery underpinning this style of enquiry rather nicely (see Howarth, 2000: 129).

A second possible route into the investigation or study of discourse may be found in attempts to identify or *recover* the intentions underpinning the language and other discursive performances of particular actors. As Josselson (2004: 5) suggests, this style of enquiry is centred on the desire to 'represent to ourselves, and the readers of our work, clearly and accurately, the message our participants are trying to convey to us'. Approached in this way, the guiding principle of research into discourse is one of fidelity to both the object and agent under study. Our task, from such a perspective, to offer a patient and faithful reconstruction of the purposes and objectives beneath the texts we are studying.

Each of the above strategies possesses obvious appeal for those attempting an engaged, critical, enquiry into specific socio-political problematics. Where efforts at discursive discovery offer a revelatory, even emancipatory, promise of truth stripped of rhetoric, quests for discursive recovery promise a similar truth of intentions and motives. My own study of textual reproductions of temporality in the War on Terror, however, departs quite considerably from each of the above analytics for reasons I would like here to make clear.

In the first instance, I am not interested in adjudicating the accuracy of any of the claims to knowing temporality made by the Bush administration in their writing of the War on Terror. Such an approach would, of course, necessitate my juxtaposing these claims to an extra-discursive reality entirely incompatible with the anti-foundational approach to the social sketched in my above discussion of discourse. For, as Foucault (1981: 67) argues in a powerful and well-known passage:

> We must not imagine that the world turns toward us a legible face which we would have only to decipher; the world is not the accomplice of our knowledge; there is no prediscursive providence which disposes the world in our favour.

A second reason for eschewing any form of discursive discovery, also implied within Foucault's citation, concerns its epistemological privileging of the observer of discourse (see also Josselson, 2004: 19). Where the aim of this style of enquiry is to uncover hidden ontological depth(s) beneath the potentially misleading surfaces of praxis and speech, a voyage of discovery of this sort would necessarily posit the existence of a privileged vantage point for observing and understanding the world. An insensitivity of this sort to our own location and (political, normative, analytical) commitments would, of course, again be problematic within the context of an approach denying the existence of any ultimate grounding for knowledge. For, as Laclau (1989: 78) suggests: 'The collapse of the myth of foundations deprives History and society of an ultimate meaning, of an absolute point of departure for political reasoning in the Cartesian sense'.

My investigation also avoids any effort to recover the intentions underpinning the Bush administration's uses of temporality within the War on Terror. Despite, again, possessing obvious normative appeal in searching for the motivations behind particular textual practices (Josselson, 2004: 13), a quest for intentionality of this sort would involve locating the origins or causes of discursive operations within the realm of the extra- or non-discursive. Such a search would not only employ a similarly problematic discursive/extra-discursive dichotomy to the above discovery-oriented approach but also tend towards an unsatisfactorily instrumentalist understanding of language. Such an approach to language as a vehicle for the satisfaction of pre-existing interests would be unsatisfactory, first, in presuming our ability to somehow access or know those intentions and, second, in obfuscating the extent to which subjectivities and interests are themselves constituted within and through discourse (see, for example, Laclau & Mouffe, 1987: 92). For these reasons, an alternative route into the analysis of discourse is necessary for our purposes: a route predicated not on discovery or recovery, but, rather, *invention*.

Recent explorations of discourse and narrative have been marked by a growing interest in the inherently, necessarily, inventive character of any reading of social texts (Desai, 1993; Caputo, 1997: 42; Patton, 2004: 29–30). Approaching the study of discourse as such a process of invention suggests that the reader of texts must abandon any search for their meaning either in authorial intentionality or in an extra-discursive realm to which privileged access has been granted. Instead, the meaning – any meaning – of a statement or practice must be approached as a contingent and precarious outcome: not only of the text under scrutiny but also of the reader's own interpretive efforts and the context in which any reading takes place. An inventionalist reading, understood thus, represents an unavoidably violent and partial in*ter*vention within a particular text: a creative 'decision' or qualified starting anew that cannot but exclude alternative readings. In this sense, a reading of this sort as with any effort to institute anything offers a circumscribed form of invention emerging from 'not the mixture but the tension between memory, fidelity, the preservation of something that has been given to us, and at the same time, heterogeneity, something absolutely new, and a break' (Derrida, 1997: 6).

This inventionalist approach to the study of discourse underpins the following discussion of temporality. Proceeding in this way, I argue, not only avoids any problematic distinction between discourse and reality of the sort discussed in the above alternative strategies but also allows my discussing the significance of particular texts within the unfolding War on Terror without searching for any agential intentionality 'behind' or 'beneath' them. Seen through this lens, it is important, of course, to recognise the violence underpinning my own reading of this conflict: to resist any temptation to speak the truth on the centrality of temporality within this particular problematic (see, for example, Agger, 1991: 113–14; Clough, 1991: 545; Lenzo, 1995: 18–19). Yet such a reading also facilitates a potentially far greater fidelity to my object of research than either of the alternatives would permit in drawing our attention to the War on Terror's aporias, heterogeneities, and even contradictions (see, for example, Norris, 1990: 21). Rather than fetishising the War on Terror as an unproblematic, bounded, and coherent totality to be embraced or rejected in its totality: to be criticised or defended through juxtaposition to an exterior set of criteria. My own reading offers an attempt to recognise and engage with its openness and to recognise the inventive character of that engagement. For, as Ashley and Walker (1990: 385) suggest in a rather different context:

> [T]o take a text seriously is to give serious attention to the unfinalized celebration of freedom and paradox that goes on within it, a celebration that the text can never really still or exclude. It is therefore to refuse to forget what a memorializing reading must forget in order to claim to retrieve a unique meaning from a text: that these classical texts are already intimately and actively caught up in a crisis of representation ... in which any supposed sovereign resolution of paradoxes of space, time and identity can never be more than a question, a problem, a paradox in its own right.

Book structure

The remainder of this book is structured around five chapters that take up the themes raised in the above discussion. In Chapter 2, 'Making Time, Shaping Time', I offer a more detailed account of the conceptual apparatus of my analysis, contextualising my research within ongoing discussions over the character and existence of temporality. Although a lot of debate has occurred concerning social science's engagement or otherwise with matters of time, I introduce in this chapter a considerable body of philosophical and sociological literature of direct relevance to our discussion. These literatures, which draw attention to temporality's intrinsically social, discursive, and historical existence, underpin my exploration into the presentation, production, and significance of claims to time within the Bush administration's writing of the War on Terror. The chapter concludes by introducing the three temporal shapes that are explored in more detail throughout the three chapters that follow: temporal discontinuity, temporal linearity, and timelessness.

Chapter 3, 'Writing Radical Discontinuity', provides the first substantively empirical chapter of the book. In this, I explore the Bush administration's persistent return to images of temporal discontinuity in the immediate aftermath of the attacks of 11 September 2001. The discussion begins by tracing the pervasive governmental positioning of 9/11 as an exceptional and unprecedented moment of rupture: a radical and traumatic interruption of comprehensible, predictable, and describable normality. This positioning, I argue, facilitated the emergence of three increasingly prominent writings of temporal discontinuity to which I then turn.

The first of these, 'Writing 9/11 as warning/lesson', relates to official discussions of those attacks as a harbinger of a radically new security environment characterised by the omnipresent threat of unconventional, undeterrable enemies. The second construction of discontinuous temporality explored in this chapter, 'Writing 9/11 as declaration/instigation of war', concerns official accounts of this event as a similar temporal interval separating now a new time of War on Terror from an old time of peace. This writing, I argue, presented the War on Terror as having already begun on 11 September 2001, irrespective of the actions and agency of the George W. Bush administration. The third, 'Writing self-transformation', relates to the less prominent, but equally significant, presentation of 9/11 as an interval separating an old time of American individualism from an entirely new time again of faith, unity, and social cohesion.

In Chapter 4, 'Writing Linear Times', I turn to the second temporal shape of the book, exploring the Bush administration's equally frequent resort to notions of linear temporality in the unfolding War on Terror. I begin this discussion by tracing the writing of 9/11 not as a unique or unprecedented event, but, rather, as the most recent example of an increasingly destructive terrorist threat. I then turn to depictions of post-9/11 America as experiencing a qualified or moderated return to normality that differs considerably from the sense of an abrupt turning away from the past considered in the final writing of the previous chapter. I conclude Chapter 4, finally, by exploring the seemingly ubiquitous proclamations of certainty, progress, and inevitable victory that accompanied the institution of this particular conflict. Approaching these writings collectively, I argue, offers a rather distinct understanding of the War on Terror's pasts, presents, and crucially futures to those considered in Chapter 3. A future, here, not of unknowable, unpredictable developments detached from any historical moorings. But, rather, one with discernible linkages to that which went before.

Chapter 5, 'Writing Timelessness', concludes the empirical discussion of the book by exploring yet another temporal shape that was equally central to the Bush administration's War on Terror. This writing of time differs considerably from those noted above, approaching temporality neither as an evolutionary pattern of gradual, incremental change nor as a revolutionary dynamic characterised by moments of rupture from which radical, qualitative, transformations take place. Instead, as we shall see, the administration was also keen to conceptualise time as an eternal pattern of return and recurrence in their inserting the War on Terror into an ahistorical, perpetual confrontation between, first, good and evil, and, second, freedom and fear. Positioning this conflict not as development from that which had previously occurred, nor as distinct from all earlier confrontations, these writings located the War on Terror as, simply, the most recent incarnation of a metaphysical, Manichean, battle.

In the concluding chapter of the book, finally, I turn to the implications of our discussion for retheorising the politics of (counter)terrorism, identity, violence, and temporality hinted at already above. Paying attention to the significance of these heterogeneous writings of temporality, I suggest, not only offers a counter-memorialising reading of the certainties that have so frequently accompanied the War on Terror's unfolding. But, also, offers considerable insights for our understanding of the socio-political far beyond these particular times of terror.

2
Making Time, Shaping Time

In Chapter 1, we encountered four reasons for taking seriously the various representations of temporality within the Bush administration's unfolding War on Terror. In the first instance, I argued that such a study would work to supplement and extend existing analyses of the relationships that exist between discourse, terrorism, and counterterrorism. By shedding light on a hitherto-neglected – and critical – aspect of these linkages, I suggested that we would be in a position to acquire a far fuller understanding of the ways in which (un)conventional violences are made both (im)possible and (il)legitimate in particular historical contexts. Second, I claimed that exploring the *different* ways in which time was constructed in this conflict would also afford us an important alternative perspective from which to engage with, and critique, this war's apparent necessity, inevitability, and coherence. Third, I argued that our discussion would also likely reveal something quite interesting about the centrality of temporal claims to reproductions of identity, security, and violence far beyond this particular historical juncture. And, finally, I noted that our study would also open space for thinking through temporality itself as an inherently, and necessarily, political phenomenon. Claims to a privileged knowledge of the movement, shape, and drivers of temporality, I argued, will always be political in their both making possible social frameworks of action and in their simultaneously militating against others.

To begin my defence of these arguments, it is necessary now to turn more explicitly to this book's primary concern: the character, existence, and functions of temporality. Attempting to do so from a social scientific background, however, is a more difficult task than we may at first imagine. This is the case, as, for some time now, academics have been debating whether matters of temporality have been afforded the attention they merit in political science, International Relations, sociology, and our other related disciplines. Although several have answered these questions in the affirmative (see, for example, Bergmann, 1992: 82; Nowotny, 1992: 423–4; Nowotny, 1994: 7; Elchardus, 1998: 35–7), others have suggested that many of the conceptual lenses, theoretical paradigms, and research questions that have historically dominated these traditions have themselves been complicit in the neglect of these matters. Lewis and Weigert's (1981: 432) effort to explore the social dynamics and dimensions of time, for example, offers a fairly typical example of the latter position, with the need for such an exploration explained by the tendency within sociology to:

> treat time as incidental to other sociological problems rather than meriting investigation in its own right. Consequently, there is a dearth of theoretical work on the sociology of time that might lead to a general paradigm for integrating the disparate empirical studies in this developing subfield. More regrettably, many sociologists do not include time as a relevant variable in their studies, or else they introduce the temporal dimensions of social organization and interaction in an *ad hoc* fashion to assist explanation of social behaviour.

From a similar perspective, almost 20 years later, Bash (2000: 190) was able to raise the very same history of non-engagement, arguing in strikingly similar language: 'The notion of time has been relatively neglected in its implications as an epistemological factor in the disciplined study of the social, and especially with regard to social theory it remains something of a forgotten dimension.'

On one level, it is of course difficult to adjudicate between these competing readings of academic history. Identifying cases of neglect, marginalisation, or exclusion necessarily incorporates a qualitative and normative judgement that extends far beyond any quantitative calculation. What I would like to do in this chapter, however, is to situate my own exploration of the War on Terror political discourse within a body of literature that emphatically *is* interested in questions of time and temporality. A body of literature, more specifically, that is interested in the social and discursive character and existence of time. By discussing the insights of these works on the main arguments of this book, this chapter seeks to further illuminate the claims raised in Chapter 1. Namely, that, first, specific governmental claims to temporality were central to the perceived coherence, legitimacy, and necessity of the unfolding War on Terror. And, second, that reflecting on this significance tells us a great deal about the relationships between temporality, identity, security, and politics more broadly.

In order to do this, I begin this discussion with a brief introduction to a concept that has attracted a considerable amount of attention in recent years: social time. This discussion offers a backdrop for our then turning to a related, yet more specific, notion of narrative time that will be of considerable use to us in the chapters that follow. I conclude this chapter, finally, by turning to the three specific temporal imaginaries that structure my own exploration into the War on Terror that comprises the remainder of this book. As pointed out in the conclusion to Chapter 1, these relate to conceptions of temporality as a radically discontinuous process; conceptions of temporality as a linear, evolutionary dynamic; and, finally, conceptions of temporality as a timeless cycle of return and revisitation.

Social time

In recent years, much of the relevant academic discussion relating to temporality has employed a distinction between something which is referred to as 'social time', on the one hand, and something quite different known variously as 'scientific', 'mathematical', 'astronomical', or 'objective' time (LaRossa, 1983: 580; Nowotny, 1992: 421–3; Elchardus, 1998: 42–3; Bash, 2000: 196; Cunliffe et al., 2004: 265–6). To better understand what is meant by the former, and to further grasp its productivity for the arguments of this book, it will be useful to first reflect briefly on the latter. And, where better to look for a notion of scientific time than within the reflections of Sir Isaac Newton, for whom 'Absolute, true, mathematical time, of itself, and from its own nature, flows equably without relation to anything external, and by another name is called duration' (cited in Turetzky, 1998: 73).

Newton's brief description of mathematical time – an entity he subsequently distinguishes from relative or common time – draws out two key assumptions that typically underpin this style of thinking about temporality as a category of scientific discourse. In the first instance, and most important for our discussion, this form of approach attributes time an ontological existence that is entirely autonomous of the social or subjective contexts of its construction. Time, thought about thus, does not come into being in the ideas, narratives, and practices of individuals or their groups and communities. It pre-dates and precedes those very ideas, discourses, and practices: existing as an autonomous fact whether or not we are aware of its existence or passage. Seen from this perspective, then, time is, quite simply, non-social, non-human, and non-discursive.

Approaching time as an objective entity of this sort suggests that we need not overly concern ourselves with the ways in which time is experienced, lived, represented, or reproduced in particular social contexts. It suggests, also, that we need not overly concern ourselves with the considerable differences that exist between distinct experiences, representations, and reproductions of time. It is enough, in this style of thinking, to approach time as a universal category to which all humans everywhere, anywhere, are equally subject. How we measure, interpret, or mark the passage of time – indeed *who* precisely engages in the measurement, conceptualisation, and marking of time – changes not its existence, nature, or impact. For, as Kroes (1984: 424) perceptively summarises: 'According to the objective theories of time flow, becoming is an aspect of physical reality, and therefore the flow of time is a legitimate physical concept which in principle can be explicated without making an appeal to "extra-physical" concepts such as "consciousness" or "mind."' Without making an appeal, either, I would add, to the social processes through which time takes on meaning at specific historical junctures.

The second key assumption underpinning Newton's conception of mathematical time relates to his emphasis on the equable flow through which this objective, universal, entity courses or moves. Conceptualised in this way, time not only exists irrespective of our perceptions, understandings, or beliefs about its existence, it also has a distinctive shape or form. Time, scientific time, comprises a succession of instants moving us constantly and continuously from past through present to future (Turetzky, 1998: 73). It is characterised by a quantifiable, measurable, and unchanging directionality: a linear, irreversible, and homogeneous directionality that progresses without pause, interruption, detour, or reversal. It is this assumption of linear directionality, of course, that explains the familiar and widespread metaphorical thinking about the passage of time as an arrow.

This approach to thinking about time as both an objective and linear category resonates with many of the everyday assumptions about temporality that many of us, I suspect, will possess. We know, on the one hand, that the future awaits us irrespective of our actions or wishes – away, of course, from the meanderings of science fiction and so forth. We know, also, that the past is gone, behind us, and done with: no matter how much we may rue, lament, or, indeed, celebrate its passing. And, it is, of course, also hard for us to imagine that this understanding of time is anything other than a universal condition of the human experience. My inability to return to the past, to project myself into the future, or to otherwise somehow control, bend, or reverse time's arrow seems not a condition of my particular spatial, historical, or social location. It is difficult to believe that my agency in these matters would differ significantly were I located elsewhere: in another place or, indeed, in another time. Once we begin to think a little more closely about our own experiences and understandings of time, however, I wonder whether the value of this strictly mathematical conception begins to appear rather more limited.

Whatever else we think we know about time, we certainly know that it is not experienced or lived in a uniform or universal manner. If we think only of the individual to begin with, my own (seemingly personal) experience of the passage of temporality is likely to differ significantly depending on my circumstances. Whether I am at work or at leisure, whether I am anticipating or dreading a future occurrence, whether I am early or late for a scheduled appointment, time – 'my' time – accelerates or decelerates in a concrete and tangible manner: a concrete and tangible manner that seems to depart quite radically from the equable flow of Newton's mathematical time. I do not experience my time as a succession of precise, homogeneous, and equivalent units (be they seconds, minutes, hours, days, months, years, or so forth), but, rather, as something that ebbs and flows according to my situation. In a very real sense, then, time – for me as an individual or agent (socially located) – seems something quite other than an abstract objectivity both beyond and above my own presence 'within' it.

If we move, very briefly, from the individual to the social, time's departure from its mathematical moorings seems to become even further pronounced – and, indeed, even more interesting. For, as we all know, different communities, groups, populations, and societies have always thought about, represented, and experienced time differently. If my own impression of the passage of temporality is (at least in part) structured by the Gregorian calendar and measured by the years that have passed since the birth of one man, I must be aware that this calendar itself has a history and context. Not only did this specific mechanism for marking and measuring time emerge at a particular historico-political juncture – as any such framework necessarily must – it has also, thus far, escaped universal imposition. And, once we begin reflecting on all of the other concepts so central to our engagements or practices of time – be they discussions of modernity; impressions of the passing of the seasons; the repetition of religious, political, or social holidays; and countless others besides – then temporality again rapidly takes on a rather different character to the mathematical conception encountered above. If time does not exist as an objective, abstract, and mathematical reality for any individual caught up in its passage, it seems to exist even less as such for the distinct pluralities of individuals distributed across the globe.

Hopefully, it should by now be becoming apparent that discussions of social time emerge out of a concern with precisely these matters that escape the more objectivist understandings with which we began. These discussions are interested, put very simply, in the ways in which time is marked, measured, and made meaningful both for and by particular individuals: particular individuals who are, of course, always already embedded in social and political contexts (see, for example, Zerubavel, 1977: 868–9). Beginning with the assumption that our understandings of temporality are central to our sense of self, our sense of others, and our sense of the places in the world occupied by both self and other, these literatures seek to reflect on and explore the particularities of specific conceptions, frameworks, and problematics of time. If my own sense of identity is bound up within particular, contingent, understandings of pasts, presents, and futures – and, indeed, particular and contingent understandings of the relations between pasts, presents, and futures – then it is absolutely crucial I acknowledge the possibility that other individuals and communities will likely frame these dynamics rather differently to me.

Of perhaps greater significance to our own discussion of the War on Terror is a second interest motivating these explorations of social time. For, not only do these accounts set out to trace the various frameworks through which time is encountered and managed (at home, at work, at leisure, and so forth), they also attempt to investigate the very real impacts posed by these frameworks for distinct individuals, groups, and communities. In this sense, discussions of social time are fundamentally interested in questions of politics and power: engaging not only with issues of management and control – *who* determines the presentation and manipulation of time in relation to particular contexts and issues? – but also with questions pertaining to the receipt or internalisation of specific temporal frames. Are particular conceptions of time encountered, for example, as taken-for-granted and natural realities? Or, are they encountered as specific socio-political frameworks of discipline and control that are themselves subject to negotiation, contest, and even resistance? As Bash (2000: 196) usefully summarises:

> Social time would refer to the temporal modalities which, integral to the reality constructions that inform everyday life in a given society,

> serve as part of an inquiry's *substantive* base – its data source. Since social time is culturally defined and socially sustained, it is patently relative and varies according to what particular aspects of which social world are under scrutiny. Thus, common-sense conceptions of periodicity, rhythms of daily routine, senses of duration, and life cadences are aspects of social time, as are manners of construing relations between past, present and future and weighting them in terms of their relative salience in everyday life.

As Bash's remarks suggest, this notion of social time provides a potentially very radical departure from the Newtonian-inspired appeals to time as an object considered above. By drawing attention to the intersubjective processes through which time becomes meaningful – or takes shape – within the lives of individuals, thinking about social time not only helps us to render the inherently contingent and constructed existence of particular temporal modalities visible as precisely that, it also demands our exploring the exclusionary and inclusionary dynamics of particular time-reckoning systems. Whose experience or understandings of time, for example, are left out within specific temporal frameworks? Which events are remembered, and which are forgotten as we forge meaningful stories of that which has gone, that which is now, and that which will be? For, as Bash (2000: 196) continues, questions such as these are all too easily passed over in our tendency towards common-sense notions of temporality:

> As members of society, interactively, engage in the social construction of time, they do so with a blithe conviction that (their) *social* time is, indeed, *real* time. As such, their notion of time tends to be invested with an evaluative dimension that lends it priority over any alternative social conceptions of time – which are then duly deprecated. In other words, temporocentrism follows as 'naturally' from the construction of *social* time as ethnocentrism attends the social construction of culture (of which social time, of course, constitutes an integral element).

To summarise, briefly, explorations of social time offer a particularly useful starting point from which to begin understanding the significance that particular temporal images and imaginaries perform within distinct socio-political contexts. Although discussions motivated by such a concern often mobilise different vocabularies, the broad interpretivist sensibility shared by these accounts promises to shed considerable light on the structuration of specific social problematics. Functionalist and quasi-functionalist analyses motivated thus, for example, have been revelatory in demonstrating the practices through which collective identities are constructed around the remembrance of specific critical events that work to bind individuals together as communities with a shared past, present, and future (Sorokin & Merton, 1937; Gross, 1985; Miller, 2000). Feminist writers have employed similar notions in tracing women's cultural and corporeal experiences of temporality as a key political strategy for exposing the inability of a strictly linear conception of time to account for all those that are deemed subject to its arrow (Kristeva, 1981; Kahn, 1989: 20–3; O'Brien, 1989: 14–16; Leccardi, 1996: 173–5; Davies, 2001: 137–8). And, considerable debate has also taken place in relation to the disciplinary implications of a fetishised 'clock time' for workers coerced into particular roles within the economies of industrialised states (compare Thompson, 1967; O'Malley, 1992; Urry, 1994; and Glennie & Thrift, 1996). Although the specific merits and conclusions of these studies need not overly concern us here, they all offer important examples of efforts to explore the significance of specific temporalised practices and existences.

From social time to narrative time

As indicated above, thinking about time as an inherently, necessarily, social phenomenon offers an analytically productive route into a fuller understanding of the construction of specific socio-political problematics. In relation to my own interest in the War on Terror, and the Bush administration's representations of temporality therein, however, it is necessary to draw on a series of more specific engagements with this broad metatheoretical insight. For, as the above discussion suggests, a concern with the social time(s) of any historical juncture connotes not only exploring productions or writings of temporality in any context, it also indicates an interest in the manner in which specific reproductions of temporality are negotiated, internalised, or read by those subject to their impacts. In this sense, exploring social time in its broadest formulation requires reflecting on the *direct* consequences or implications of particular temporal narratives within the lived experience of individuals and groups.

The preceding remarks relating to the connotations of social time point to one important reason for my turning to a more specific – yet equally anti-essentialist – account of temporality in the following exploration of the War on Terror's discursive construction. In the first instance, while an investigation into the persuasiveness or otherwise of particular claims to temporality within this context would likely shed considerable light on its particular structure and form, such an investigation would also throw up considerable epistemological difficulties for our analysis. As noted in the first chapter above, where it is remarkably difficult to search for any directionality linking motives or intentions to the productions of texts, so equally difficult obstacles emerge when searching for the direct consequences of claims to temporality. How would we access the minds of those encountering specific (constructed) pasts, presents, and futures within the War on Terror? How would we adjudicate whether the Bush administration's writings of rupture, continuity, or timelessness were being met with approval or suspicion; whether they were being internalised by their audiences or, indeed, resisted? In light of these rather difficult questions, we are on grounds considerably more secure if we limit our analysis to the productions and framings of time alone: rather than embarking on a reflection into their receipt and readings. And, for this reason, I want to turn now to my understanding of narrative time as a particular way of engaging with the productive (ontological) insights of social time, while avoiding their (epistemological) difficulties.

The notion of narrative time offers an attempt to draw attention to the human construction and expression of temporality within and through narrative. If we take narrative very broadly as 'an account of events occurring over time' (Bruner, 1991: 6), explorations of narrative time set out to explore, first, which events are privileged as significant within particular accounts of pasts, presents, and futures. And, second, the ways in which these events are related to one another in the forging of ostensibly meaningful, ostensibly coherent, stories and plots (see, for example, Ricoeur, 1980: 171; Ricoeur, 1990). More accurately, perhaps, discussions of narrative time are concerned with exploring *how* specific events are inscribed with meaning or significance *through* their temporal ordering and relations to others. For, although the world we encounter throws up all manner of discrete events one after the other, it is the connections we make between them that positions specific occurrences as something other than, simply, a list or a series. The patterns that we forge or impose on events – be they assertions of continuity, discontinuity, repetition, and so forth – are of the utmost significance in our attempts to make sense of the worlds into which we are thrown. And those patterns, of course, may always be thought about otherwise. For, as White (1980: 27) suggests:

> The notion that sequences of real events possess the formal attributes of the stories we tell about imaginary events could only have its

> origins in wishes, daydreams, reveries. Does the world really present itself to perception in the form of well-made stories, with central subjects, proper beginnings, middles and ends, and a coherence that permits us to see 'the end' in every beginning? Or does it present itself more in the forms that the annals and chronicle suggest, either as mere sequence without beginning or end, or as sequences of beginnings that only terminate and never conclude?

As White's remarks indicate, explorations of narrative time are interested in the various ways in which events are sequenced or connected. This is the case, for it is the very act of sequencing and connecting events in specific – non-predetermined ways – that attributes them coherence and meaning for particular individuals and communities (see also Bearman et al., 1999: 502). From such a perspective, there are two key issues to which the observer or student of narratives must remain attentive when exploring the emergence of specific temporalities within narrative practices. In the first instance, there is the chronological or episodic dimension of a narrative: that which works to identify those occurrences that are pertinent or relevant to the tale being told – be that a tale of trauma, tragedy, conquest, and so forth. And, in the second, there is the configurational or emplotted dimension of a narrative: that which enables the fashioning or contrivance of those significant, yet potentially dissociable, events into the wholes that make up our well-made stories with beginnings, middles, and ends (Ricoeur, 1980: 178; Ashley, 1989: 263–4). Although we need remain sensitive to the (violent) interpretive work that accompanies the identification of these significant wholes (compare Smith, 1980; Chatman, 1981; Cronon, 1992: 1368–9; Kreiswirth, 2000: 311–14), the discursive inventionalist approach sketched in the previous chapter enables us to do this quite comfortably.

This brief introduction to the concept of narrative time is intended to shed light on two key insights that it poses for the chapters that follow. In the first instance, by tying the meaning and significance of any event to its narrated relations with other occurrences, this analytical tool will help move us beyond essentialising any of the entities with which we are interested in our discussion of the War on Terror. Whether we are focusing on 9/11, America(ns), terrorism, or, indeed, the War on Terror as a totality, this approach will enable our taking seriously the contingent, constructed – and hence political – existence of each of these phenomena. That 9/11, for example, was inserted into markedly different narrative times by the same administration figures speaking, effectively, at the same time, should indicate that this event's status and import was not an inevitable product of the attacks themselves. Put otherwise, the framings of any of these entities within specific conceptions of temporality, I will argue, must be approached not as peripheral to their underlying extra-discursive reality. But, rather, these framings must be approached as constitutive of those entities' (temporal, social, and political) position and, indeed, their significance.

In the second instance, if the emplotment of any coherent narrative derives from the violent and partial attempt to link particular historical entities into coherent wholes, we must always remain aware to the whole host of other (potentially relevant) occurrences that will necessarily have been excluded for the constructed story to make sense. As we shall see in the chapters that follow, discussing 9/11 with reference to Pearl Harbor, for example, offered a markedly different narrative interpretation of those attacks than would one referencing the daily suffering of individuals through global patterns of structural violence. By drawing attention to the exclusionary structuration of these claims to temporality – claims to temporality that were absolutely essential to the War on Terror's coherence, necessity, and legitimacy – then, this concept of narrative time also helps focus our attention on the second sense of the political outlined in the previous chapter.

Temporal shapes

To conclude this chapter, I wish finally to introduce the three radically different ways in which the Bush administration represented or narrated temporality in their writing of the War on Terror political discourse. Where each of the following chapters takes one of these as its object, it will be valuable to be clear here on the distinctions that exist between these temporal imaginaries. To this end, following Graham's (1997) heuristic typology (see Johnson, 2001: 90), I distinguish between three broad temporal 'shapes' that were all central to the construction of this conflict. These shapes refer to claims to radical discontinuity, claims to temporal linearity, and claims to timelessness.

Radical discontinuity

Conceptions of radical discontinuity begin with the premise that the movement of time represents a punctuated, disjunctive, dynamic. In contrast to evolutionary notions of historical transformation which presuppose both continuity and discontinuity within their object of study, conceptions of radical discontinuity posit moments of absolute historical break separating qualitatively discrete temporal periods (Graham, 1997: 112–14). Where the most simplistic versions of Enlightenment thinking attribute a sense of smooth linear progress or teleology to history's course, discontinuous readings of our pasts, presents, and futures locate radically distinct eras or epochs sharing little or nothing with one another (Nowotny, 1992: 435–7; Richardson, L., 2006: 603). In this sense, constructions or narratives of radical discontinuity typically encompass three key elements. The first is a claim to relative stability, coherence, or uniformity within a particular object at a specific historical juncture. The second is the occurrence of a significant, frequently novel, event that acts as a rupture or 'interval' (Fortier, 1999: 47) both interrupting the hitherto stable configuration and bringing something else into being. And, the third, finally, is a period of new stability or coherence following the interval: a period of new stability, importantly, that is qualitatively and entirely distinct from the original situation.

Presented in such stark terms, it is perhaps unsurprisingly difficult to identify many uncritical accounts of radical discontinuity within the humanities and social sciences. The complexities of the social world invariably resist the appeal to homogeny and delimitation required of this first temporal shape. With that caveat in mind, however, it is certainly possible to approach Kuhn's (1996) conception of scientific paradigms and revolutions, and Foucault's (2002: xxiv, although see also, Foucault, 1990: 100) archaeological inquiry into the Western episteme's 'wholesale transformations' of the eighteenth and nineteenth centuries as sharing something of a concern with historical discontinuities: as sharing something of a concern with the significance of historical discontinuities that are not fully capable of capture by appeals to quantitative, linear, transformations in science and so forth (see Palmeri, 1999). And, at a more general level, we can, of course, also discern notions of radical, qualitative discontinuity within the less nuanced uses of some of the more familiar binary oppositions that continue to pervade social scientific discourse today: be they modern/postmodern, Cold War/Post-Cold War, Fordism/Post-Fordism, Theory/post-Theory, and many others besides. For, underpinning each of these examples is the suggestion that a relatively uniform, unitary historical period has been interrupted and subsequently replaced by a new, and distinct, historical period.

Chapter 3 takes this first temporal shape as its focus, exploring three distinct examples of its persistent employment by the Bush administration in their writing of the War on Terror. Each of these constructions,

I argue, worked to position 9/11 as a temporal interval separating two incommensurable historical periods. And, as outlined therein, this positioning also worked, importantly, to produce those attacks as significant because of this status. Signalling the conclusion of an old (pre-9/11) time characterised by relative security; the absence of war; and a self-interested, individualistic American public, those attacks were depicted as, simultaneously, ushering in a radically new time characterised by the coexistence of irrational and dangerous enemies; a necessary new type of war; and an increasingly united, pious, and reflective America. This contingent and exclusionary positioning of 9/11, I argue, posed profound and discernible implications for the unfolding War on Terror, including, most notably: working to legitimise the Bush administration's efforts to respond to the threat of terrorism through unprecedented, exceptional, strategic mechanisms; disciplining social and political opposition to this newly enacted war; and facilitating the emergence of specific collective identities – both self and other – that were central to this war's construction and enactment.

Linear times

The second broad temporal shape that we explore in the discussion that follows concerns governmental representations of linear time(s) within the emergent War on Terror. In contrast to the above conceptions of radical discontinuity with their emphasis on uniqueness, rupture, and abrupt qualitative change, notions of linear time approach the unfolding of history as a process of gradual, incremental, transformation. Conceptualised in this way, temporality possesses a traceable, knowable, and potentially even predictable directionality, with time unfolding as a structured, irreversible, dynamic of connected beginnings, middles, and ends (see, for example, Robbins, 2001: 529–31). In this sense, linear time may be approached as inherently chronological: concerned with the straightforward ordering of events from past through present to future – with each of these entities replacing the last in an ordered and orderly dynamic.

As is well known, linear conceptions of time as a cumulative, directional dynamic draw on a long and vigorously contested history within Western social and political thought. In political science, for example, evolutionary theories of social change (compare John, 1999: 43–5; Kerr, 2002), notions of path dependency (see, Thelen, 1999: 384–8; Mahoney, 2000; Alexander, 2001: 253–60), and teleological accounts of the 'end of history' (Fukuyama, 1992, although see also Fukuyama, 1995: 31–5), all mobilise an effectively linear conception of temporality, as do earlier, less critical, renditions of 'modernization theory' (see Joas, 1999: 468). Although these approaches differ markedly in their conception of the agents and drivers of temporal change, and although advocates of these approaches typically reject simplistic approaches to history as a straightforwardly unidirectional process, each conceptualises change as a gradual, transformatory dynamic within which earlier events or decisions constrain or structure subsequent occurrences. Within sociology and the philosophy of history, similarly, explicitly normative notions of progress and historical decline also attribute an overarching directionality to temporality: understanding this process as either one of continuous advancement in the object under study (Graham, 1997: 45–82), or one of incremental degeneration from a superior, 'better', time of the past (Gross, 1985: 56; Graham, 1997: 83–111). In each of these examples, a coherent focus on notions of sequence, causality, and determination traverses the understanding of temporality on offer.

In Chapter 4, I explore three distinct writings of linear time that were again returned to repeatedly throughout the unfolding War on Terror by key representatives of the Bush administration. I begin by considering presentations of 9/11 and the continuing terrorist threat as an evolutionary development from earlier terrorist attacks on America(ns) and others. This writing, I argue, explicitly problematised the notions of radical discontinuity introduced above: locating 9/11 not as a temporal interval separating two qualitatively distinct times of security. But, rather, as one moment of progression within a broad, identifiable and cumulative politico-historical trajectory of increasingly deadly, increasingly destructive, unconventional assaults. Second, I turn to governmental presentations of the aftermath of 9/11 as a gradual return to a relative, moderated, normality within the United States. In common with the third discontinuous writing considered above, I approach this writing as an important mechanism for depicting those attacks as a bringer of change within American society. In contrast to that construction, however, I also suggest that this writing conceptualised those changes not as a radical departure from the past, but, rather, as a series of modified transformations that were necessary in order to leave the 'essence' or 'core' of that collective self unbroken. We conclude our exploration in Chapter 4 by exploring the seeming ubiquity of constructions of certainty, predictability, and knowable progress within the unfolding War on Terror itself. These claims to the inevitability of future victory against terror, I argue, again ascribed a cumulative, linear directionality to the movement of temporality: locating key moments in the Bush administration's new war as steps on a pathway to a desirable end. And, as with the discussion of radical discontinuity in Chapter 3, I explore a number of significant politico-discursive implications of each of these linear constructions in this exploration. Claims to a unidirectional movement of time, I suggest, were integral to the writing of political certainty into this war, to the legitimisation of potentially exceptional security mechanisms within this conflict, and to the production, again, of specific collective identities around which this war was framed by the administration.

Timelessness

The final broad construction of temporality explored in this book relates to governmental discussions of time as a recurrent, cyclical, and paradoxically 'timeless' dynamic in the unfolding War on Terror. As with the conceptions of linear time considered above, notions of temporal recurrence have a long history within Western thought. Graham (1997: 141), for example, locates such a conception in Lucretius's first-century-BC reflections on the 'waxing and waning' of cultural groupings, while Brockmeier (1997: 191), similarly, argues that the idea of 'circular time' has been found in all civilisations that conceptualise humanity as part of the natural world. In the social and political sciences, more recently, prominent examples of this approach to the temporal include Kennedy's (1991:487) conceptualisation of the 'rise and fall' of great powers as 'a story of replication, of imitation, of attempts to catch up, by countries and societies eager to draw level with – or eclipse – another nation that, at the time, is in the lead' (see also Layne, 1993); neo-realist or structuralist conceptions of 'hegemonic war' as a recurrent feature of an unchanging anarchical international system (Gilpin, 1988); and Modelski's (1978) 'long cycles' of global politics. At a more microlevel of analysis, explorations of daily work routines in the lives of individuals also demonstrate this concern with recurrence (Alheit, 1994).

In contrast to the above approaches to temporality as either a linear or discontinuous process, conceptions of timelessness emphasise the persistence of repetition and return, sameness and equivalence, between chronologically distinct historical eras. As Thucydides (cited in Starr, 1966: 27) suggested, the study of history is for 'those who want to understand clearly the events which happened in the past and which (human nature being what it is) will at some time or other, and in much the same ways, be repeated in the future'. In this sense, notions of cyclical temporality neither prioritise the emergent or the novel around which conceptions of radical discontinuity are constructed, nor do they share the concern with unidirectionality or irreversibility evident in the linear approaches noted above. Instead, while positing the existence of historical change, and possible differences between chronological periods, conceptions of timelessness indicate that transformations are neither radical nor entirely unprecedented. Rather, change occurs as a continuous and patterned circular dynamic, where one temporal cycle may be meaningfully compared to another. As McFarland (1991: 258) bluntly summarises: 'As time passes, things go back to where they were before'.

The final substantive chapter of our discussion takes as its object two particularly prominent writings of timelessness that were again invoked with great frequency by the Bush administration. The discussion begins by exploring the framing of this war as the most recent moment in a continuous historical battle between good and evil. It then turns to a secularised version of this construction of timelessness, locating the persistent representation of this conflict as a repetition of earlier conflicts between freedom and fear. Each of these writings, I argue, offers a further fascinating challenge to both the linear and discontinuous conceptions of 9/11, the War on Terror, America(ns), and so forth introduced above: presenting this conflict as qualitatively equivalent to prior historical struggles against fascism, communism, and, indeed, others. Depicted in this way, the present historical juncture was to be understood by observers and participants as not exceptional or unique. But, rather, as effectively an echo or reappearance of continuous temporal patterns. We conclude Chapter 5 by approaching these two constructions as of great import once more to governmental writings of this conflict. Most importantly, as we shall see, these appeals to timelessness worked to inscribe normative certainty into this conflict – presenting it not only as necessary but also as just.

Conclusion

We opened this book by noting the absolute saturation of the Bush administration's unfolding War on Terror with claims to a privileged knowledge of temporality. As outlined in Chapter 1, our discussion here offers one attempt to take this saturation seriously by tracing the emergence, structuration, and implications of these governmental efforts to speak the truth about time in this context. To more fully sketch the specific approach to temporality underpinning this analysis, this chapter has introduced a variety of literatures that ground my own reading of these *Times of Terror*. Rather than viewing time as an objective, extra-discursive object of knowledge that exists in isolation from its social perception or construction, the chapters that follow take the movement, dynamics, and drivers of time as inherently, necessarily, human productions. Put simply, time does not exist apart or aside from our social and narrative understandings of pasts, presents, futures, and their relations. It is the very act of forging and (de)linking these pasts, presents, and futures that brings time into being as a meaningful entity. Moreover, it is in the act of forging these pasts, presents, and futures into temporal shapes that specific subjects and objects of time are themselves constituted as meaningful. For, as we shall now see in Chapter 3, there is a great deal of import attached not only to the ways in which particular events and identities are positioned as significant within and through efforts at imagining or constructing the temporal, but also to the implications that follow these (contingent and violent) acts of imagination and construction.

3
Writing Radical Discontinuity

We concluded the previous chapter by introducing the three temporal shapes that structure our discussion of the War on Terror: radical discontinuity, temporal linearity, and timelessness. As outlined there, this chapter takes the first of those shapes as its object of analysis by exploring the Bush administration's persistent return to notions of temporal interruption and radical change in their writing of this conflict. To this end, we begin by exploring the multifaceted, and seemingly ubiquitous, representation of 9/11 as an absolutely or relatively unprecedented occurrence. Drawing on linguistic constructions of temporal, discursive, and normative rupture, as well as non-linguistic commemorative practices, I argue that the Bush administration coherently positioned those events as a radical departure from an imaginary antecedent horizon of expectable, comprehensible, describable, and acceptable normality. As such, and in contrast to much of the other relevant literature here, I do not approach those events as an initially traumatic void awaiting narrative filling. Rather, I approach those attacks *as narrated as* an initially traumatic, indescribable, void awaiting possible understanding with the distance of time. This positioning of 9/11 as an interruption of temporal, discursive, and normative normality, I argue, was not only important for inscribing some form of singular significance into those events. It was also crucial for establishing the discursive terrain upon which that significance could be projected forwards into the future in subsequent reproductions of those attacks as a moment of radical temporal discontinuity or transformation.

The second part of this chapter explores three related narrative strategies through which the Bush administration did indeed project 9/11's (constructed) exceptionality into the future. Approached collectively, I argue that these narratives worked to both sediment and extend the attacks' ostensibly destabilising impact: constituting 9/11 as a path-breaking, world-changing, historical event. Positioned thus as a defining moment of American, indeed global, existence, these narratives created the impression that life would never return – could never return –to pre-9/11 normality. The certainties of the past had not only been interrupted that day, they had been irrevocably, and forever, lost.

The first of these writings relates to the persistent official representation of 9/11 as a warning or lesson of present and future insecurity. Building upon positionings of the attacks as an unforeseeable bolt from the blue, this writing projected this powerful sense of unpredictability forwards as indicative of a new and pervasive existential condition. By representing the attacks thus, I argue that the Bush administration successfully constituted 11 September as a temporal interval separating an old time of manageable and predictable security from a new time of unpredictable enemies and permanent threat. Although a particularly productive strategy for legitimating specific – equally exceptional – responses to 9/11, I argue that this narrative also functioned as a powerful interpellative call for Americans to recognise and share in a new, emergent, collective identity.

The second writing that interests me in this chapter relates to constructions of 9/11 as a declaration or instigation of a fundamentally new war. Positioned as a mythical originary moment bringing the War on Terror into being, this narrative reaffirmed the attacks' status as a temporal interval dividing two qualitatively distinct historical periods. Separating now not only an old time of security from a new time of threat, 9/11 was also here represented as distinguishing an old time of peace from a new time of (new) war. As outlined below, this narrative offered a particularly powerful and productive discursive resource for responding (aggressively, militarily) to the attacks of 11 September 2001, not only in its predication on a claim to first victimhood but also in its assertions that a new conflict with the forces of terror(ism) had been entered that day.

The third section of this chapter explores the related, yet again distinct, official representation of 9/11 as a moment of radical self-transformation within the United States' populace. Once more marking the movement between two incommensurable historical periods, this writing also produced those attacks as a separating moment: a separating moment here dividing an old time of selfish, materialistic individualism from a new time of collective reflection, piety, and moral clarity. This narrative was interesting, I suggest, not only as an alternative interpellative mechanism through which some measure of agency and control could be (re)introduced into the 'shocked' and 'traumatised' American public. But also for introducing an explicit normative ambiguity into the 9/11 attacks: locating positive consequences of those occurrences alongside their more immediate (constructed) 'devastation', 'destruction', and 'trauma'.

In the concluding section of this discussion, I tie the above three narratives to my interest in the relationships that exist between temporality, identity, violence, and politics. In the first instance, I argue that the Bush administration's fondness for claims to temporal discontinuity offers a powerful demonstration of the productivity of this form of theorising temporality for both legitimising and delegitimising specific violences. In the second, I argue that the centrality of these claims to the constitution of particular collective subjectivities similarly indicates the linkages that exist between time and identity. And, finally, that each of these relations came into being – had to come into being – through the exclusion of alternative readings of pasts, presents, and futures offers evidence, I suggest, for the claim that temporality is inherently, necessarily, political in the sense with which we began our discussion in the preceding chapter.

Prefacing radical discontinuity: Writing 9/11 as rupture

If the purpose of this chapter is to understand the Bush administration's successful positioning of 9/11 as a moment of radical historical discontinuity, we need first to engage with representations of those events as a moment of temporal rupture. Where assertions to any form of temporal discontinuity are made possible by the identification of a qualitative movement from an old time to a new, then locating something of a break with the past represents a necessary (if not sufficient) prior condition of that movement. Necessary, because without such a break pasts, presents, and futures emerge as connected with one another in a manner connoting not rupture but continuity. Insufficient, because there is always the possibility that identified ruptures may be (re)positioned as mere anomalies, curious abnormalities, or temporary interruptions of broader temporal dynamics. And, as we shall see in the chapters that follow, indeed, the Bush administration did explicitly attempt such a renegotiation of 9/11's exceptionality by simultaneously, counter-intuitively, locating those events within very different temporal imaginaries to that with which we are here concerned.

Motivated thus, this section traces four key ways in which the Bush administration represented the September 11 attacks as a moment of abnormality or break. It begins by exploring explicitly temporal constructions of rupture in significations of those events by their date, writings of 9/11 as an unprecedented and unpredictable event, and discussions of those attacks as a traumatic loss of innocence. Each of these writings, I argue, produced those events as a sudden and unexpected break with an imaginary horizon of stable, predictable normality. I then turn to a distinct yet related construction of discursive abnormality, exploring the persistent and paradoxical writing of the attacks as somehow resistant to language and interpretation. This writing, I argue, again constituted this event as a sudden, traumatic break from the normal: producing September 11 as here so exceptional or unique as to have rendered our previously adequate structures of description simply inadequate to the capture of this exceptionality. In the third instance, I briefly explore the administration's explicitly normative claims to 9/11's abnormality, noting the frequent invocation of tropes of tragedy and barbarism and the like in the aftermath of this event. And, finally, I turn to the ways in which specific commemorative and memorial practices worked to reaffirm the attacks' rupturing impact.

Writing temporal rupture

On 13 September 2001, the United States Attorney General John Ashcroft addressed a press conference called to reflect on the now (in)famous events of two days prior. He began his remarks to that gathering with what appeared an entirely appropriate note of solemnity, stating, quite simply, 'September 11 2001 was a day of unspeakable violence and outrage' (13/9/2001). At Freedom Hall, New York, a little under three weeks later, President Bush (3/10/2001) offered a related reflection on the attacks' domestic impact, noting, 'there's no question that the economy has been affected by September 11th.' While his 13 October expression of support for Tom Ridge as the newly instituted head of Homeland Security included the following assertion of Ridge's capacity for the job: 'Governor Ridge ... knows what we're up against, because his own state was one of the three where Americans died on September 11th' (Bush, 13/10/2001).

In one sense, the above statements clearly emphasise different aspects of the 9/11 attacks and their immediate aftermath in tracing their human, economic, and strategic-political impacts. If we approach them collectively, however, they offer a striking illustration of one of the ways in which those events were written into a distinctive temporal imaginary: a distinctive temporal imaginary predicated on a notion of rupture or break. They do so, of course, by demonstrating the explicit discursive linkage that was forged by the Bush administration (and others) between the four aircraft crashes, on the one hand, and, on the other hand, the date upon which those crashes occurred. From a (curiously full and formalised) marker of contextual information in Ashcroft's 13 September denunciation, this date – 11 September 2001 – was rapidly truncated and transformed into a signifier – a name – for the attacks themselves.

As Derrida (2003: 85–94) argued in an early interview following those events, to fully understand the sense of exceptional significance that 9/11 rapidly accrued, it is imperative that we take this apparently innocuous act of naming quite seriously. We must do so, because this metonymical borrowing of a date to stand in for an event worked as a powerful strategy for transferring the singularity of the former unto the latter. If 11 September 2001 marks a unique historical occasion: a specific and distinct moment in time incapable of direct repetition or precedent, then the employment of this signifier to identify just one occurrence that day functioned to inscribe a similar distinctiveness into the attacks. By so frequently and forcefully tying those events to their unique temporal mooring, then, the Bush administration (re)created those events as themselves equally unique. And, as noted above, claims to temporal rupture and discontinuity are predicated on the identification of such unprecedented historical moments.

If the naming of the 11 September attacks by their date inscribed those events with an implicit singularity, this inscription was supplemented, importantly, by a number of more explicit attempts to construct them in this rather specific manner. In the first instance, we encountered a whole series of direct assertions to those attacks' unique and exceptional character: whether framed normatively – as in Bush's (16/9/2001) claim that 'we've never seen this kind of evil before' – or strategically – as in Ashcroft's subsequent (25/7/2002) reference to this 'unprecedented assault'. As Vice President Cheney (18/10/2001) summarised rather uncompromisingly, 'September 11 was a day like no other we have ever experienced.'

In the second, we witnessed the persistent effort to represent 9/11 as so unique that the attacks had necessarily defied any possible prior prediction of their occurrence. As the President himself argued on several occasions, 'None of us could have envisioned the barbaric acts of these terrorists' (15/9/2001), for, 'They struck in way that was unimaginable' (26/9/2001a). A theme that was subsequently picked up by the Deputy Secretary of Defense Paul Wolfowitz in his statement to the House and Senate Armed Services committee: 'Until three weeks ago, an attack like the one we suffered September 11th seemed unimaginable to most Americans' (3–4/10/2001).

And, in the third, we also encountered the persistent, more graphic, writing of 9/11 as a traumatic loss of innocence. Although invoked rather more rarely than the tropes considered above, this writing once again reaffirmed the impression that we had witnessed a unique and exceptional break with the past. Where Secretary of State Colin Powell's (19/11/2001) remarks at the University of Louisville, Kentucky, offered a distinctly individualised framing of this loss: 'Others of us merely lost our innocence. We can never look at a jetliner flying in a clear blue sky the same way again', others such as Richard Haas and Condoleeza Rice depicted this transformation as a truly collective experience:

> So, on September 11, our innocence ended, and we entered the post-post-Cold War world, a period when increasingly potent transnational challenges intersect with still important traditional concerns. The attacks were a grim reminder of how the march of globalization has raised the stakes from transnational threats.
>
> (Haas, 22/4/2002)

> It's going to take years to understand the long-term effects of September 11th. But even now we are beginning to recognize that there are certain verities that September 11th reinforced and brought home to us in the most vivid way. ... there has been an end to innocence about international politics and about our own vulnerability. We see that wars of consequence are not mere relics of a bygone era.
>
> (Rice, 29/4/2002)

If we take the above statements together, we begin to get some sense of the pervasiveness with which the Bush administration represented 9/11 as a rupture of temporal normality. Positioned, repeatedly, as an unprecedented, unpredictable, traumatic loss of innocence, the events swiftly emerged as an exceptional interruption of expectable, foreseeable, ordinary existence. Written as a definitive, irreversible, break from the familiar, the events were reduced, quite simply, to a 'shock' (Ashcroft, 28/1/2002), a 'surprise' (Bush, 7/12/2001). Not predictable or foreseeable, much less preventable, 9/11 appeared as a sudden, abrupt, bringer of chaos. An unprecedented abruptness, of course, that worked to inscribe import into those events by virtue of their sheer exceptionality. Yet, as we shall see in the following section, this production of singular significance was further reaffirmed in the widespread, and seemingly paradoxical, framing of those events around a related series of claims to their discursive exteriority.

Writing discursive rupture

Consider the following discussions of 9/11 taken from remarks made by Attorney General John Ashcroft, Secretary of Defense Donald Rumsfeld, and President Bush on 21 September, 15 November, and 18 April 2002 respectively:

> The only word to describe this is the word 'indescribable'.
>
> (Ashcroft et al., 21/9/2001)

> Today, America is once again at war – a war every bit as dangerous and threatening to our national survival as that war that so tested America's courage and resolve. And again, America's citizens have met the challenge. Order had been maintained, laws have been obeyed, and love has prevailed over unspeakable evil and destruction.
>
> (Rumsfeld, 15/11/2001)

> Our Nation has come to realize the tragic toll that crime takes, and we have developed the resources to ease crime's physical, emotional, and financial impact. This support network, which was already in place on September 11, made us better prepared to deal with the unspeakable pain and tragedy inflicted by the terrorist attacks.
>
> (Bush, 18/4/2002)

From even a cursory glance at these statements, a clear resonance emerges with the above attempts to position 9/11 as an interruption of the expectable. Once again, 9/11 is here depicted as something exceptional: something of a departure from that constituting the normal. And, once again, 9/11 emerges as significant *because of* this departure from that which constitutes the normal. Yet there is a distinction to be made in these two articulatory moments where these statements refer not (only) to a break from a hitherto-stable horizon of the predictable, but, rather (or also), to a break from a hitherto-stable horizon of the representable. What these statements offer, put otherwise, is a vivid illustration of the administration's persistent and continuing writing of 9/11 as an event so unique that it had not only shocked and surprised those bearing witness to the attacks but also ruptured or broken the adequacies underpinning our very structures of language and speech.

So abnormal were those attacks, in these writings, that our existing systems of representation simply could not capture or tame their singularity. So abnormal were those attacks, in these writings, that a language capable of describing them simply did not exist. A similar construction of radical exteriority, then, to that offered above. Yet a construction predicated on the positing of a distinct (if related) antecedent field of discourse.

If the above statements demonstrate the Bush administration's positioning of 9/11 beyond the realm of representation, that particular framing was complemented by governmental accounts stressing the attacks' similar resistance to any form of legitimate, meaningful, sense of understanding. In President Bush's 17 September response to a query concerning the inevitability of retaliatory war, for example, he stated quite bluntly, 'They slit the throats of women on airplanes in order to achieve an objective that is beyond comprehension' (Bush, 17/9/2001a). The State Department's Coordinator for Counterterrorism Francis X. Taylor appeared similarly perplexed in his remarks to the Committee on International Relations, suggesting: 'There is no excuse, no justification, no rationalization for these acts of mass murder against innocent people' (Taylor, 25/9/2001). For Donald Rumsfeld (11/10/2001), addressing the Pentagon Memorial Service one month after the attacks, those events simply represented, 'the mystery of tragedy'. While President Bush (11/10/2001a), at the same commemorative event, argued even more powerfully, even more personally, 'we know that if you lost a son or daughter here, or a husband, or a wife, or a mom or dad, life will never again be as it was. The loss was sudden, and hard, and permanent. So difficult to explain. So difficult to accept.'

My interest in these above claims to incomprehensibility lies with their similitude to all of the accounts of the 9/11 attacks we have been exploring thus far. Although heterogeneous in the posited reason for the attacks' defiance of comprehension – varying here between their rationale, totality, and temporal abruptness – they together function as a collective reaffirmation of 9/11's exceptionality. Positioning those events still further as a revelatory rupture of the normal – a revelatory rupture of the limits, here, of comprehension – these claims to their meaninglessness worked to (re)confirm their status as an interruption or breakage with a world that could be grasped. For, as written above, the discursive failure engendered by 9/11 lay not with those subjects struggling, yet failing, to understand what had occurred on that morning. The failure resided instead in the very uniqueness of that which those subjects had witnessed. From such a perspective, then, these claims to incomprehensibility offered yet another contribution to the reproduction of those attacks as an interruption or breakage from a hitherto-adequate, stable, horizon of normality. Escaping not only prediction and foresight, 9/11 here, now, also escaped both language and cognition.

Writing normative rupture

The above discussion has attempted, thus far, to illustrate the Bush administration's persistent attempt to position 9/11 somewhere beyond the realm of expectable, predictable, describable, *and* comprehensible normality. This composite positioning, I have suggested, (re)produced those attacks as a coherent and significant rupture of the familiar in framing those attacks as a moment of such singular exceptionality as to have exploded the limits of our taken-for-granted assumptions and experiences concerning the normal. Yet there is one further theme that links the above statements upon which I would like briefly to reflect in concluding my discussion of these linguistic representations of rupture. That theme, most evident in the above condemnations of 9/11 as a moment of murder, evil and tragedy, relates to the ubiquitous writing of those attacks as yet another form of exteriority. It relates to the writing of 9/11 as a moment of *normative* exteriority.

In President Bush's (11/9/2001a) first public announcement of 11 September 2001, he referred to the (then still unfolding) attacks as a 'national tragedy'. Where Richard Jackson (2005: 31–8) has already identified the administration's persistent return to this trope throughout the following months, I would like to reflect here on the power and functions of this particular framing as it relates to our interest in temporality more specifically. Notions of tragedy, and the related language of 'horror' (Cheney, 29/11/2001), 'disaster' (Bush, 12/9/2001a), and 'calamity' (Ashcroft & Mueller, 21/9/2001), resonate, of course, with the framings we have considered thus far. Yet there is a slight difference of emphasis or tone that separates this vocabulary from the above accounts: a difference that revolves around these claims' explicitly normative denunciation of the attacks. If constructions of unspeakability, for example, evoke images of exceptional destruction and pain, constructions of tragedy supplement these images with a rather more direct claim to unmitigated, unwarranted victimhood. Not only did 9/11 represent 'the hour of our direst need' in these normatively oriented framings (Wolfowitz, 31/1/2002), it represented an hour of need brought forth entirely without justification or desert. A terrible interruption of untragic normality neither foreseen nor merited by the American populace or their leaders: a horrific rupture of yet another field of the ordinary.

This persistent return to the language of tragedy and disaster offered a powerful discursive resource in its own right. By positioning 9/11 as a sudden, catastrophic, interruption of innocent lives, it allowed the administration's continual and repeated condemnation of those attacks as a pure, unthinkable, abhorrence. If we turn to some of the more graphic and detailed accounts of this tragedy that were offered in the attacks' aftermath, however, the provocative character of this grammar becomes clearer still. Not only were we subject to a series of symbols and signs stressing those events' resistance to justification, we were subject as well to a series of extended reflections on their direct and indirect effects that left no doubt at all as to their awful, tragic, unfolding. The President's 13 September remarks following his hospital visit to survivors of 9/11 offer a very clear example of the evocative character of these framings:

> Some of the folks could talk, and they described the horror of the incident, the moment. They talked about escaping, going through fire, crawling through debris. It was clear that they were fighting for their survival then, and like every patient up there, they're still fighting for survival. And it was just a sobering moment for Laura and me. But we again thank the hospital, the docs, the nurses and, of course, again tell the families that the nation prays for those who have been injured by this unbelievable act of terror.
>
> (Bush, 13/9/2001a)

Bush's account masquerades as an attempt to relay the 9/11 experience for those directly caught up in the explosions. Concretising (Lazar & Lazar, 2004: 232) the ordeal faced by the victims crawling for survival through fire and debris, he leaves no ambiguity as to the unbelievable, corporeal, horror encountered by those directly impacted that day. By casting those victims simultaneously as *fighters* possessed with sufficient courage and resolve to overcome their hellish environment, Bush's depiction also neatly introduces the twinning of victimhood and heroism that was so central to this writing of normative rupture. If America(ns) had been attacked without justification, they had met those attacks with strength and determination. As his subsequent, 7 October remarks, at the 20th Annual National Fallen Fire Fighters Memorial Tribute suggest, this merger of strength and innocence could not but assist in the (re)casting of those events as an exceptional occurrence:

> Thank you very much. Joe, thank you for those kind words, and thank you for your outstanding service in a difficult time for our

> great land. ... Two years ago this weekend, I attended a memorial ceremony in New York City honoring fallen firefighters. And standing nearby were Chief Peter Ganci and many others who are now gone. None of us on that day could have imagined what was to come, the scale of the emergency, the enormity of the danger, the magnitude of the evil. Yet, each one of those firefighters felt a strong calling and knew its risks. On September 11th, that calling led them into burning towers on a mission of rescue. Within a single hour, more than 300 firefighters were lost. And our nation still mourns. They did not live to know who had caused the destruction, or why. They only knew their duty. And that was to go in, to follow the faintest cry, to search for the trapped and helpless, and to save those who could be saved. A few days ago, one New Yorker described firefighters as 'the kinds of guys you look up to.' Every one of you here knew exactly what he meant. The courage and loss we saw in New York is found in every community that has laid a firefighter to rest. Hardly a week passes in America when a career or a volunteer firefighter does not fall in the line of duty.
>
> (Bush, 7/10/2001)

I have cited the President's remarks to this Tribute at length because they offer a particularly interesting example of this positioning of 9/11 as a moment of tragic abnormality. Although in many ways similar to his 13 September remarks, this later account moves subtly, almost imperceptibly, between two distinct, yet related, referents of that tragedy. In the first instance, at the beginning of his statement, Bush clearly returns to his earlier construction of national emergency in framing 9/11 as a collective disaster for America(ns) as a totality: 'a difficult time for our great land', in this example. As his discussion continues, however, the President turns to a much more direct and localised depiction of the attacks' impact on a very specific sub-community of firefighters that were affected unduly by the horrors unfolding. In this sense, then, although employing the same grammar of sudden danger and death to his earlier remarks, this apparently seamless movement between witnesses and participants works to reproduce 9/11 as a collectively, if differentially, shared tragic experience. By personalising this experience through invoking the unexpected absence of Chief Peter Ganci, Bush also here emphasises the very real, very human, impact of the attacks on American *people*. And, as Jackson (2005: 35) has argued, this rhetorical strategy of personification was a particularly powerful mechanism for eliciting empathy, and one, of course, employed frequently, in early official discussions of 9/11 (see, for example, Bush, 15/10/2001; Ashcroft, 24/1/2002).

Beyond Bush's attempt to negotiate two distinct subjectivities in his above reflections on the consequences of 9/11, there are two further reasons that motivate my interest in reading his remarks here quite carefully. First, in comparison to any of the statements considered thus far, this passage offers a far more explicit attempt to write those attacks around images of heroism and courage. From their immediate aftermath onwards, discussions of 9/11 as a multifaceted moment of heroic endeavour were never very far from the administration's reflections on the attacks' impact and significance. Whether, as in the above statement, by reference to the courageous efforts of firefighters and other first responders risking their personal safety so that others may live (see, for example, Bush, 14/10/2001; Powell, 29/11/2001; Rumsfeld & Myers, 11/12/2001). Or, through invocations of the ultimately unsuccessful efforts of those much-discussed passengers travelling aboard United Airlines Flight 93 (see, for example, Bush, 15/9/2001; Cheney, 16/9/2001). 9/11 was persistently scripted as a moment of desperation *and* valour: an occasion of goodness as well as tragic destruction. And, as is hopefully clear, these constructions of courage and heroism again positioned those events beyond the realm of the normal, the realm of the familiar.

Second, Bush's remarks to the American firefighting community are also intriguing in their here introducing a first moment of ambiguity into the 9/11 attacks themselves. Where this chapter thus far has concentrated on the positioning of those events as an entirely singular, unprecedented occurrence, this particular occasion offered an illuminating demonstration of the precarious, contestable, construction of this positioning. For, as he suggests at the end of his statement, 'The courage and loss we saw in New York is found in every community that has laid a firefighter to rest. Hardly a week passes in America when a career or a volunteer firefighter does not fall in the line of duty.' If the rhetoric of heroism, tragic victimhood, and sudden death worked to situate those events beyond the realm of the normal, what we have here is an explicit recognition that these claims to exceptionality or singular significance could themselves have been quite plausibly contested. Although still framing those attacks here as unquestionably memorable, Bush's audience on this particular occasion worked to temper this framing somewhat. Not, now, fully unprecedented, nor fully singular, 9/11's status appears here already a little less stable than we had been lead to believe.

It would be remiss of me to leave my discussion here without drawing attention to the surfeit of more directly condemnatory reflections on September 11 that also characterised the Bush administration's unfolding War on Terror. Although I will have more to say of these denunciations in the chapters that follow, the persistent reproduction of those attacks as heinous acts of violence (Ashcroft, 11/9/2001; Thompson, 12/3/2002), cowardly assaults (Bush, 11/9/2001b; Powell, 11/9/2001), and evil or despicable enactments of terror (Bush, 11/9/2001c; Bush, 7/10/2001; Rice, 16/10/2001; Ashcroft, 26/11/2001), again, clearly, positioned 9/11 as a rupture of something. Juxtaposed, here, to an imaginary horizon of defensible, acceptable, and normatively legitimate behaviour, those events were once more located as a radical exteriority interrupting the normal. Complemented by discussions of barbarism (Bush et al., 15/9/2001; Rumsfeld et al., 11/9/2001), atrocity (Bush & Megawati, 19/9/2001; Bush, 10/10/2001a), abhorrence (Ashcroft & Mueller, 20/9/2001), and murder (Powell, 23/9/2001; Cheney, 18/10/2001; Ashcroft, 6/12/2001) 9/11's status as an absolutely indefensible act became very quickly solidified. In Ashcroft's (14/9/2002) framing of particular relevance to our discussion, the events were produced, simply and starkly, as an 'evil that ruptured the peace of that morning'.

Remembering rupture

At the beginning of this book I introduced my understanding of the War on Terror as constituting a specific and contingent political discourse. Drawing on Ernesto Laclau's conception of discursive structures, I argued that it was important to avoid positing any form of ontological distinction between the linguistic and non-linguistic moments of this historical problematic. For, although analytical separable, the War on Terror's language and practices only become meaningful when we approach these disparate moments as part of a limited, and incomplete, system or totality. The language of this conflict only makes sense when it is viewed in relation to the objects it purports to refer to. And, those objects or practices become fully comprehensible only when viewed through the lens of the language that both justified their existence and called them into being. If I am interested in representations of time in this particular context, then, I am ultimately interested in the ways in which representations of time both constituted and made possible this conflict's material enactment.

With this note in mind, it is appropriate here to finish my description of the (re)production of 9/11 as temporal rupture by exploring some of the ways in which the Bush administration performed or remembered this rupture across non-linguistic domains. Although memories, of course, may be forged or constructed in the language we use, they are also sedimented or embedded in social consciousness by a whole range of practices designed to breathe life into that which has passed. That these practices actively work to produce those objects they claim to reflect has been long noted by theorists of social memory and time. And, in the aftermath of the 9/11 attacks we witnessed a not-insignificant array of governmental commemorative practices that served to remind us, yet again, of their exceptionality.

As Jenny Edkins (2003: 103) has noted, the official commemoration of the 9/11 attacks began remarkably rapidly after their commencement. Before two of the four aircraft had even reached their final destination, President Bush (11/9/2001b) had already called for a minute's silence to remember the events, asking 'the American people to join me in saying a thanks for all the folks who have been fighting hard to rescue our fellow citizens and to join me in saying a prayer for the victims and their families'. In making this early appeal to public commemoration, Bush's remarks opened the door to a veritable surfeit of subsequent practices designed also to bear witness and remember this unspeakable tragedy. And, as the months passed, a whole series of further traditional or 'conventional memory projects' (Wagner-Pacifici, 1996) were repeatedly instituted and observed: projects including, but not limited to, further minutes of silence, memorial services (Bush, 13/9/2001b; Bush, 14/9/2001a), and the half-mast flying of the United States flag (Bush, 12/9/2001b, 14/9/2001b).

These rather familiar efforts at public remembrance and grief tap into long-standing practices of commemorating moments of profound historical significance. As such, they were important in reaffirming 9/11's significance as a disruptive, exceptional event that we have been exploring in our discussion thus far. By revisiting, replaying, those attacks and their consequences, they sought to remind onlookers that we had indeed witnessed something of genuine significance that day. To remind us that we had witnessed something of significance even if we did not yet fully know how to articulate or make sense of those events. Even more interesting for our present concern, however, was the institution of a number of more 'anomalous' or unusual memory projects (Wagner-Pacifici, 1996) designed not only to reflect the exceptionality of 9/11 but, designed, also to somehow mirror that exceptionality. The designation of all future September 11s as Patriot Days (Bush, 4/9/2002), for example, worked to preserve this particular date as an unforgettably unique site of traumatic collective memory. The official Pentagon memorial of the attacks, attempted a similar function, in seeking 'to permanently record and express the sheer magnitude of loss through an architectural experience of a place radically different than what we encounter in our daily lives' (Pentagon Memorial Project, Undated). On 11 March 2002, six months after the attacks, we witnessed the equally unusual unveiling of the 'September 11 postage stamp' (Bush, 11/3/2002b) reaffirming, once more, those events meriting a singular public commemoration. And the institution, finally, of the '9/11 heroes medal of valor' awards ceremony 'to honor those public safety officers who gave their lives on September 11' (Bush, 9/9/2005) performed a similar function, in reperforming, once more, the attacks' exceptionality within a similarly unusual memory practice.

There already exist a number of excellent academic studies that have set out to explore various commemorations or (re)mediations of 9/11 across a range of discursive spaces and sites (see, for example, Silberstein, 2002; Butler, 2006; Simpson, 2006). And, it is beyond the scope and purpose of this book to retrace that ground here with a detailed semiotic reading of the memorial performances noted above. What I would like to stress here, however, is the role that these performances played in (re)articulating, (re)writing, 9/11 as a moment of both abnormality and rupture. An event so exceptional, so exterior to the field of the normal, that to refuse to remember it would have been simply intolerable. And, if the more conventional commemorative performances drew an implicit, unsatisfactory, equivalence between those attacks and other tragic occurrences, those more exceptional representations of 9/11's exceptionality at least promised to help us confront its uniqueness. For these reasons, these memory practices offered a final crucial element in this writing 9/11 as a moment of temporal rupture, for, as Colin Powell (10/9/2002) stated at the September 11: Bearing Witness to History Exhibit:

> It is important to the American people that this moving exhibit of objects, sights, sounds and stories from September 11th should be displayed here among the collections that constitute our national memory. It is also important that people all over the world understand how deeply September 11th is seared into the fabric of our national life, a fact symbolized by the flag which so gallantly draped the Pentagon's shattered wall and which will have a place of honor in our ceremony today.

The politics of rupture

Much of the language and other practices explored in the above section will already be familiar to those who watched with interest as the War on Terror unfolded. What I have attempted to demonstrate here, however, is the collective import of these practices for the constitution of 9/11 around a very particular conception of temporality. In the first instance, through marking those events by the date of their occurrence *and* locating 9/11 in a grammar of unprecedented singularity, the Bush administration quite explicitly positioned the attacks as a unique historical experience. Deemed to lack precedent, parallel, or any form of legitimate comparison, September 11 was produced, simply, as a moment sui generis. An exceptional episode in American, perhaps global, history: the likes of which had simply never before been seen. An exceptional interruption had happened that morning, then, of an implicit, imaginary, horizon of normality.

In the second instance, once we begin to take the other above practices seriously, as I think we must, we begin to get a fuller picture of the pervasiveness and power of this particular, temporal, positioning. Descriptions of the 9/11 attacks as somehow escaping description and understanding quite clearly reinforced and complemented the Bush administration's explicit claims to historical exceptionality. By representing those events as so disruptive, so exterior to our expectations of the normal, the Bush administration consistently told us that we simply did not possess the adequate linguistic and cognitive resources to make sense of that we had witnessed. Claims to the attacks' normative abnormality in discussions of tragedy, horror, murder, evil, and the like compounded this sense of shock and astonishment: reproducing 9/11 once more as a radical abnormality. And the hurried, yet continuing, rush to commemorate whatever had occurred on that day worked to remind us, once again, of that occurrence's distance from that which we could – that which we should – reasonably expect. In short, all of these practices together worked to position 9/11 as an interruption or absolute break from an implicit realm of expectable, predictable, describable, comprehensible, acceptable, and indeed forgettable, normality. They positioned this event as a rupture of the normal.

As suggested above, claims to rupture or historical exceptionality do not in themselves constitute a narrative of temporal discontinuity. Conceptions of time as a discontinuous, broken, dynamic necessitate some reflection on that which comes after a break as well as that which has been broken. That said, there are two reasons for our taking this composite writing of rupture quite seriously at the start of this exploration of our first temporal shape. In the first instance, this writing was absolutely fundamental to the three discontinuous narratives I wish to consider below. That the Bush administration was able to successfully position 9/11 as a moment of American, indeed global, transformation depended entirely upon these prior claims to 9/11's exceptional character. If radical, rather than mere quantitative, change can only be predicated on a sense of departure from the normal, the above writings provided that sense in abundance. In the second instance, and more narrowly perhaps, the above writings were also important where many of the ostensibly exceptional characteristics attributed to 9/11 therein were themselves projected forwards as symptomatic of the new time, the new era, that we were subsequently informed had arrived so unexpectedly that day. To offer one example, the unpredictable, unforeseen status of the attacks considered above, was itself appropriated as indicative of a new security environment comprising similarly unpredictable enemies and threats. In short, if this broad writing of rupture facilitated the extension of 9/11's significance into the future, the specific indicators of rupture structuring this writing were themselves already giving shape to the future we were awaiting.

Before turning to the three explicit narratives of temporal discontinuity that comprise the remainder of this chapter, it is important here to reflect briefly on the political and its relation to this specific writing of rupture. For, if the political concerns both the constitution and exclusion of specific social realities, then this particular positioning of 9/11 offered abundant evidence of both. In the first instance, by framing those events as a moment of rupture or break, this composite writing clearly militated against alternative, equally plausible, conceptions of 9/11. In the second, this writing also posed quite considerable implications for the shape and legitimacy of the War on Terror itself that was instituted, subsequently, in response to those events. Let us consider each of these in turn.

At the risk of stating the obvious, the Bush administration's persistent, pervasive, effort to position 9/11 as an unprecedented, and entirely unpredictable, break with the past was predicated upon a rather curious reading of contemporary world history. Reduced to a sudden, traumatic, moment of shock, this writing worked to isolate those attacks from any number of potentially relevant backdrops that could have been quite legitimately invoked to understand this occurrence. With a long-running history of antagonism between the United States and al-Qaeda erased at a stroke, earlier violences in East Africa, America, and far beyond (Jackson 2005: 16, 98) were conspicuous by their absence in this writing of rupture. And, if we take into account Osama bin Laden's well-known and continuing hostility to US Foreign Policy that was similarly missing from the above story, we arrive at a still fuller sense of the precariousness of this positioning. How different would the 9/11 attacks have appeared had these historical contexts been posited? Would not the above claims to exceptionality have seemed rather less solid?

While an explicit acknowledgement of these historical contexts would have afforded a very different flavour to our understanding of 9/11, this story of tragic exceptionality was also made possible by a rather more serious exclusion. Where specific dimensions of those attacks (such as, perhaps, their logistical conduct) may indeed have appeared unprecedented, a rather presumptive and partial leap of faith was required to translate these dimensions into the radically singular violence connoted by the above framing. As we have known for a long time, we need to be very careful in reducing the world's violences to those loud bangs and blasts that capture the attention of publics and policymakers alike (Galtung, 1969; Galtung & Hoivik, 1971). For, the most pernicious, and destructive, bringers of human suffering are typically those author-less tragedies that pass us by largely unnoticed with depressing familiarity. Read against such an alternative backdrop of broader, structural, violence, 9/11's singularity would, again, have been increasingly difficult to defend. For, As Stanley (2002: 208–9) persuasively notes:

> [T]he *New Internationalist* illustrated the point that on September 11 (assuming annual deaths were evenly spread) 24,000 people died of hunger, 6,020 children were killed by diarrhoea and 2,700 children were killed by measles. The latter 'mundane' and preventable deaths, related to the structural implementation of Western 'liberal democratic' agendas, do not serve to delineate a hierarchy of pain or 'victimhood': rather they allow us to understand that, in some instances, the distinctions between 'perpetrators' and 'victims' overlap.

An even more provocative attempt to draw attention to the implications of this singular framing was evident in the email message circulated by a Brazilian anarchist organisation shortly after the attacks:

> 35,615 children died out of starvation on September 11, 2001.
> victims: 35, 615 children (source: FAO)
> where: poor countries
> special tv programs: none
> newspaper articles: none
> messages from the president: none

> solidarity acts: none
> minutes of silence: none
> victims mournings: none
> organized forums: none
> pope messages: none
> stock exchanges: didn't care
> euro: kept its way
> alert level: zero
> army mobilization: none
> conspiracy theories: none
> main suspects: rich countries. (Cited in Collins, 2002: 168–169).

In positing these efforts to contextualise 9/11 rather differently, I should be clear that I have no interest in engaging in a reductive 'calculus of suffering' here (Klein, 2002: 148). Trading the misery and pain of particular communities against one another would be as distasteful as it would be, ultimately, counterproductive. What I hope to have illustrated, however, is the very precarious character of this specific (temporalised) writing of rupture. For, by situating those events differently, as these alternative readings emphatically attempt to do, the above claims to exceptionality become very quickly rather less self-evident, very quickly rather less concrete, than the Bush administration appeared to want to have us believe. If 9/11 was, indeed, a 'mysterious tragedy' as the Bush administration sought to remind us so frequently, the uniqueness of this mystery seems rather more contestable than the above constructions of exceptionality would indicate.

I have already suggested that this writing of singular tragedy worked to inscribe political and historical significance into that which has become known as 9/11. By coherently positioning those attacks as an exceptional, unprecedented, event, the Bush administration clearly attributed that occurrence far greater import than an alternative framing of the sort offered by Klein and Collins would have permitted. Although it is interesting to note that this inscription of significance could, indeed, have been otherwise, there is a second, more urgent, reason for our subjecting this particular framing to critique. If we consider the uses to which this construction of singular significance was subsequently put, we very quickly come to see that this writing was integral in preparing the foundations for the quite extraordinary range of security responses that followed those events. For, as Bush's (7/6/2002) reflections on the Homeland Security project demonstrated quite vividly, this temporalised claim to self-evident importance was absolutely central to the subsequent institution of the War on Terror itself: 'I knew I needed to act right after September the 11th. ... We needed to do something very quickly right after the attack.'

As Bush's comments rather forcefully demonstrate, this framing of 9/11 as a disruptive break with the normal offered a remarkably powerful justification for responding to those attacks. Positioned as so unquestionably critical that he simply 'knew he needed to act' in retaliation, this (contingent, contestable) writing of rupture effectively set in place the entire discursive apparatus that was, eventually, to become known as the War on Terror. If it did not, by itself, structure the specific and far-reaching array of politico-military responses that followed in the wake of that day of violence, this production of exceptionality certainly worked to decontest the need for his administration's doing *something*. So self-evident was the attacks' significance in this writing, that to have done nothing would have appeared simply indefensible. And, with the need for some form of action thus established, it was but a short step towards declaring war (and everything that accompanied it) to which we will turn in the following sections.

Beyond its legitimatory function in preparing the space for instituting the subsequent War on Terror, this writing of rupture also quite clearly assisted in absolving the Bush administration from any responsibility in their failure to prevent those attacks (see, for example, Borradori, 2003: 148). By positioning 9/11 as so unique an event that we simply could not have foreseen its occurrence, any form of meaningful, sustained, discussion over governmental culpability was instantly, magically, erased at a stroke. With the 9/11 Commission's subsequent performance of procedural reflection only reminding us that so little had been done prior to the events of 11 September, the attacks effectively, paradoxically, functioned as a rallying cry for *supporting* the Bush administration. Had they been positioned rather differently – as a security failure, for example – then the administration's sustained programme of military interventions and homeland security would have been far more difficult to defend. By writing 9/11 as a moment of unavoidable exceptionality, then, this production of rupture actually worked to (re)inscribe confidence into an administration that had been previously, and widely, deemed incompetent for the task of running the country (Silberstein, 2002: 39). If there is nothing new in the production of crises for mobilising political support, we must remain here attentive to the contestable foundations on which this particular framing was founded.

A third consequence that followed this positioning of 9/11 as historical break related to its effective delegitimisation of any calls for a fuller, more patient, effort to understand or explain the attacks and their causes. Pacified by appeals to 9/11's incomprehensible, mysterious nature, those bearing witness to the violence that day were asked, simply, to trade any sustained discussion of their origins for the more comfortable, stupefying, language of shock and evil. As Condoleezza Rice (10/4/2002) remarked at the National Commemoration of the Days of Remembrance:

> [R]ight now, all of us are enduring a time of testing, loss, and fear; a time when our vulnerability to evil and the certainty of our mortality are all too clear; a time when once again our intellect is insufficient to answer the question, 'Why?' And at these times more than ever, we are reminded that it is a privilege to struggle for good against evil.

As Rice's remarks vividly demonstrate, this comforting yet reductive appeal to tropes of surprise, trauma, and rupture translated directly into a systematic refusal of any sustained analysis of the attacks. Although it did appear permissible to ask why 'they' hated 'us' in the events' aftermath, this desire to allow hatred perform the explanatory work for onlookers only deferred the attacks' mystery one further step. Notions of hatred – as with notions of evil, and uniqueness – offer no genuine space for the political. They stand in as a cover, a mask, hiding potentially tangible – indeed, potentially legitimate – material grievances (see, for example, Rabinow, 2001). And, as several have noted, this closure of understanding or enquiry from the attacks' status, import, and reasons not only seriously delimited the administration's capacity to prevent future violences, it also prepared the grounds for the subsequent and rapid decline of global support for their War on Terror campaign that had been initially instituted amidst a whole raft of declarations of empathy (see, for example, An-Na'im, 2002: 168; Parekh, 2002: 270–6).

Finally, in writing September 11 as a rupture of normative normality specifically – as a permanent loss of innocence for the American public – this particular temporal framing also functioned as a powerful inscription of blamelessness and goodness into the immediate *and* collective victims of the attacks. Constituting this event as a shared abnormality whose unjustified impact and consequences far transcended those killed or wounded that day, this writing offered a powerful interpellative call to the American public to recognise and participate in their collective opposition to those capable of understanding, justifying, and conducting such an abhorrence. Although we return to this Manichean, antagonistic, framing of global political life in Chapter 5, it is worth noting its existence already in this first temporal shape. For, as noted in our introductory discussion, claims to temporality are intimately bound up with identity politics. If 9/11 represented a rupture, 9/11 also represented America's rupture.

To summarise briefly, we have seen now the swiftness and pervasiveness with which the Bush administration wrote 9/11 into a very specific conception of temporality. Reproduced persistently as an unprecedented, somehow indescribable, tragedy, those attacks were persuasively positioned as an emphatic, chaotic, rupture of the normal. Even if we were unable to fully describe, to fully understand, what had happened that day, we somehow already knew that we had witnessed something significant. And, because of that world-changing, unprecedented, significance, we also knew already that some form of response would soon be forthcoming. By tracing some of the exclusionary practices that made this writing appear plausible for those subject to it, and by exploring the implications it posed for subsequent events, we have already embarked on our concern with the relationship between time and the political. For, alternative framings of those attacks as not rupture but, rather, repetition or continuity would quite clearly have challenged the coherence of this construction and its subsequent consequences. And, as we shall see in the chapters that follow, such writings were indeed offered by the Bush administration in their creation of this war.

Writing temporal discontinuity

With the benefit of hindsight, both the ease and the speed with which Bush's administration positioned 9/11 as a break with the normal may seem a little surprising. The sense of assuredness, of absolute confidence, with which the past was identified as being ruptured that day appears now, looking back, rather immodest, if not, frankly, extraordinary. Perhaps even more surprising, however, was the rapidity with which these claims to 9/11 as a moment of rupture were translated into something of even greater historical significance in the equally widespread positioning of 9/11 as a bringer of temporal *discontinuity*. Presented, then, not only, as a departure from that we had known, from that we had taken for granted as expectably, describably, understandably, normal, this framing (re)positioned those events as a *permanent*, irreversible, departure from a past that could never be recaptured.

On 16 September 2001, Tim Russert interviewed Vice President Cheney on NBC's weekly *Meet the Press* television programme. Following some relatively predictable remarks concerning the seriousness of the issues tabled for discussion that morning, the Vice President immediately turned to a theme that seemed thoroughly to capture the administration's imagination in the following months. He did so by claiming, with complete conviction, 'I think the important thing here, Tim, is for people to understand that, you know, things have changed since last Tuesday. The world shifted in some respects' (Cheney, 16/9/2001). As the weeks and months progressed and we witnessed the launch of a new War on Terror, Cheney warmed to this theme of a global shift having recently occurred. Speaking, repeatedly, of the emergence of a 'new normalcy' (Cheney, 25/10/2001; 14/11/2001; see also Ridge et al. 2/11/2001) that day, the Vice President appeared emphatic in his conviction that life was not any longer as it once was: whether in the United States alone or indeed in the world beyond its borders. Yet he was far from alone in this attempt to locate 9/11 as an irreversible, permanent, temporal disjuncture, with the President, for example, telling Congress of his impression of night falling on 'a different world' (Bush, 20/9/2001) on 20 September. From the Department of Justice, Deputy Attorney General Larry Thompson (2/11/2001, 8/11/2001) described the attacks and their aftermath as 'a watershed in the history of the United States', with his senior, John Ashcroft, referring similarly to 'a defining moment in the history of our Nation' (24/1/2002), 'a unique moment in American history' (7/10/2002), and a 'post-September 11' environment or world on multiple occasions (14/11/2001, 16/9/2002, 1/10/2002): most poetically, perhaps, in stating, 'On September 11, 2001, the wheel of history turned and the world will never be the same' (Ashcroft, 25/10/2001). From the Department of Defense, the attacks were presented in related language as a historical 'turning point' (Wolfowitz, 14/11/2001; Feith, 21/4/2002), the aftermath of that point, simply 'not normal times' (Rumsfeld & Wolfowitz, 12/12/2001). While, in the National Security Advisor Condoleezza Rice's (9/10/2001) more extended account:

> September 11th was one of those rare dates that forever divides history into distinct categories of before and after. Everyone will remember what he or she was doing on September 11th. It was a day when the dark and impossible became a horrific reality for our country and for the world. We commonly hear the refrain that everything changed on September 11th. In many ways that is true.

As each of these claims neatly demonstrates, the Bush administration was incredibly keen to locate 9/11 as a far more significant event, even, than the interruption of normality discussed already above. Written in this way, as 'the day that forever changed our country' (Bush, 6/6/2002), the attacks were constituted not only as an (absolutely or relatively) unprecedented departure from that which had been previously expected or witnessed, but, more important, they signalled a moment of transition or change from an old time to a new. In this sense, these statements offer an indication of the persistent governmental attempt to produce 9/11 as a coherent temporal interval or dividing point separating two qualitatively discrete historical eras. Of their persistent attempt to locate those attacks as a moment of radical temporal discontinuity ushering in a present/future so distinct from the past.

As noted already, the remainder of this chapter is concerned with exploring three distinct, yet related, governmental efforts to expand on these assertions of temporal discontinuity. To do so, it begins by tracing the writing of those attacks as a warning or lesson that we had recently entered a new time of radical insecurity. Upon this, we then turn to representations of 9/11 as a bringer of (new) War on Terror, before concluding here with an exploration of claims to the self-transformatory impact of those attacks. If we take these three writings of 9/11 together, I shall argue, it is possible to identify the emergence of a very particular, persuasive, and powerful temporal imaginary within the Bush administration's construction of this unfolding war.

Writing 9/11 as warning/lesson

The first writing of temporal discontinuity I wish to discuss emerged from the Bush administration's persistent positioning of 9/11 as a moment of entry into a new, vastly more dangerous, historical period. Initially structured around a reading of those events as a warning of the imminent possibility of future terrorist attacks, this construction was subsequently, and importantly, expanded in presentations of 9/11 as a less precisely determined warning of present and future insecurity. Written in this way, not only had we entered a world in which our (new) terrorist enemies would look to visit ever more violence on us at any moment, we had entered a world of additional, unknowable, future security challenges: the likes of which had never been witnessed before. Let us take each of these in turn.

Representations of 9/11 as an indicator of possible future terrorist attacks emerged very quickly in this event's aftermath. Building on the above constructions of 9/11 as an unpredictable, unforeseen, interruption of normality, the attacks were positioned here as something of a warning or wakeup call (Wolfowitz, 14/11/2001, 31/5/2002). A warning or wakeup call alerting policymakers and publics alike to the existence of a new and dangerously unconventional threat to our existence. As Colin Powell, perhaps the administration's least alarmist of personnel, suggested in a 14 September press briefing: 'I think it is also fair to say that the events of the 11th of September have fundamentally changed the way in which people look at terrorism and acts of terrorism' (Powell, 14/9/2001). His subsequent, 23 September, 'Meet the Press' interview returned to this theme:

> It's deeply moving to know that, one, we had this kind of vulnerability and that there were people out there who we knew were out there but never really had a sense of how determined they were to strike us in this way. And it means that as we go forward, we will have to work harder to protect ourselves, work harder to find this kind of enemy, work harder to defend ourselves.
>
> (Powell, 23/9/2001)

Powell's remarks quite explicitly position 9/11 as an indicator of a then-recent qualitative shift. Although hinting at some underlying transformation in the strategies and resolve of this ostensibly new terrorist threat, the primary focus of his writing here concerns the attacks' impact on the administration's knowledge or understanding of terrorism. As he subsequently suggested on the eve of the attacks' mensiversary, 'the events of September 11 brought home to us in tragic fashion the global reach of terrorists in today's world' (Powell, 10/10/2001), for, as Larry Thompson (2/11/2001, 8/11/2001) subsequently argued, 'How we as a nation and a government think about our vulnerabilities to a terrorist attack is forever changed'. In John Ashcroft's (14/9/2002) understanding, similarly, 9/11 also functioned as a subjective, lagged, marker of a transformation that had already taken place in contemporary terrorist capabilities, where, 'September 11 alerted all of us to our vulnerabilities.' While Paul Wolfowitz's (2/12/2002) response to a question on Iraqi weapons systems indicates both the continuing pervasiveness and subsequent significance of this particular understanding of the 9/11 attacks:

> I think it goes back to some, an understanding, of just how profoundly the events of September 11th changed America's understanding of the risk and the stakes. I think indeed if we had understood that it was possible for 3,000 Americans to die in a single day, and have the

> graphic experience of what that entailed, we would probably have taken much more forceful [action] against Afghanistan long before. But we can't wait until 30,000 Americans or 300,000 or even possible 3000,000 die as a result of an attack by weapons of mass destruction to deal with the threat posed by countries that have the weapons and develop them and support and work with terrorists. That is the heart of the issue.

As Wolfowitz's remarks suggest, this writing of temporal discontinuity structured 9/11 as a moment of realisation within which our new, contemporary climate of insecurity had been suddenly, perhaps finally, grasped. Although an important construction in its own right, this presentation was supplemented, significantly, by an even more dynamic account of those events' agency. This supplementary presentation positioned the attacks not (only) as a 'wakeup call' to an already-existent transformation in terrorist violence. But rather (or also) as an active bringer of qualitative change: an active bringer of change marking the *arrival* of a fundamentally new terrorist threat. Although these related constructions were often interwoven into one another in governmental discussions, John Ashcroft's 2 October media availability session appeared to ascribe greater significance to the attacks than that within the above accounts of a shifting awareness, with his claim: 'I believe that additional terrorist acts are possible. And I believe the kind of attack which we endured shows that the risks of such possibilities are substantial' (Ashcroft & MacAulay, 2/10/2001). If similar to those first representations, this writing presented the attacks themselves as a *demonstration* of new risk: rather than as a facilitator of a belated recognition of a threat that was already present on 11 September 2001.

Ashcroft returned to this construction of 9/11 as an agent of change numerous times over the following months. On 26 February 2002, for example, he argued that 'America and the world have been awakened to a new threat from an old evil – terrorism' (26/2/2002, see also, Ashcroft, 18/5/2002; Ashcroft & Ridge, 10/9/2002): combining this production of rupture, interestingly, with a more explicit attempt at historical contextualisation than any of the presentations we have encountered thus far. From the White House, the Vice President's 25 October address to the Republican Governors Association positioned the attacks likewise, with his suggestion, 'The enemy has shown the capacity to inflict great damage on the United States, and the only safe way for us to proceed is to assume there will be more attacks' (25/10/2001, see also, Cheney, 23/10/2001, 21/2/2002), while the President himself returned repeatedly to this writing for some time beyond 9/11's immediate aftermath (see, for example, Bush, 6/6/2002, 11/6/2002). And, from the State Department, finally, Colin Powell (16/9/2001, 11/7/2002) was equally keen to recall the experience of 9/11 in his calls for increased vigilance against the possibility of future terrorist attacks. Within each of these discussions, then, the significance of the September 11 attacks extends even beyond the bounded parameters of the (unprecedented, unpredictable, indescribable) event constituted by the representations discussed already above. Positioned as a warning that terrorist actors possessing both the capabilities *and* determination to attack the United States had either emerged on, or could no longer be ignored following, that day, 9/11 was coherently located as a temporal interval separating a new time of insecurity and threat from an (implicit) old time of relative, predictable security. The vulnerabilities and danger so graphically, so tragically, demonstrated in those hitherto-unprecedented attacks had not simply passed with their conclusion. Rather, further such incidents conducted by the same protagonists or others like them remained a very real risk that could be ignored or neglected no longer.

Before turning to the distinctively political character of this writing of 9/11 as temporal discontinuity, it will be valuable here to reflect on one further extension of this construction that again imbued the attacks with a significance stretching beyond their own singularity. Presenting those attacks not (only) as an indicator or bringer of new terrorist threats, but, far more seriously, as a warning of future unforeseeable security challenges, this writing became increasingly visible as attention shifted more persistently to the possibility of confrontation with Saddam Hussein's Iraq. Yet, as Wolfowitz's above-noted remarks to the House and Senate Armed Services Committee on 3–4 October 2001 suggest, this construction was again evident from the attacks' early aftermath:

> [A]s we prepare for the battles ahead, we must recognise that these strikes were not just an act of war – they were a window into our future: a future where new enemies visit violence on us in startling ways; a future in which our cities are among the battlefields and our people are among the targets; a future in which more and more adversaries will possess the capability to bring war to the American homeland; a future where the old methods of deterrence are no longer sufficient – and new strategies and capabilities are needed to ensure peace and security; these attacks were an assault on our people and our way of life; but they were also a wake-up call – one

> that we ignore at our peril ... the September 11th strikes caught us by surprise. We must prepare ourselves for the virtual certainty that we will be surprised again.
>
> (Wolfowitz, 3–4/10/2001)

Wolfowitz's discussion offers a particularly vivid example of the positioning of 9/11 as a radically discontinuous break within the United States' security environment. Drawing on the invocation of surprise introduced in the above discussion of rupture, Wolfowitz expands this possibility of future surprises to an existential, omnipresent, objective condition: the then-Deputy Secretary of Defense here seemingly capable of recognising the 'virtual certainty' of equally unpredictable attacks less than a month after September 11's (unpredicted) occurrence. In this extended presentation, then, the lesson or 'wake-up call' of 9/11 was not that America need prepare for equivalent attacks by a specific, identifiable, enemy. Rather, that the United States had been catapulted suddenly into a world of ambiguous, nameless, and radically unforeseeable threats: a world of 'known-unknowns' in Donald Rumsfeld's (Rumsfeld & Myers, 12/2/2002; Rumsfeld, 6/6/2002) memorable phraseology. By explicitly rejecting 'the old methods of deterrence' as an obsolete and anachronistic tool with which to confront this new world of threat, Wolfowitz clearly (if implicitly) juxtaposes this post-September 11 era to an imaginary Cold War era of relative stability and predictable enemies underwritten and made possible by rationalist models of deterrence. As Edwards (2004: 157) has noted, 'One of the first sentiments to emerge from the rubble was the idea that 11 September represented some sort of global dividing point, a transition from an orderly past – as seen, that is, in the rosy glow of hindsight – to a present (and future) now fraught with uncertainty and menace.'

For evidence of the discursive continuity that this expanded claim to 9/11's transformatory impact subsequently enjoyed, we could do far worse than turn to Donald Rumsfeld's 31 January 2002 remarks at the National Defense University in Washington, DC. These remarks merit citing at length for the clarity with which the Secretary of Defense positioned 9/11 as a 'global dividing point' of the sort noted by Edwards. As his reflections make clear, once more the attacks emerge here as an interval separating two entirely heterogeneous security regimes:

> Preparing for the future will require us to think differently and develop the kinds of forces and capabilities that can adapt quickly to new challenges and to unexpected circumstances. An ability to

> adapt will be critical in a world where surprise and uncertainty are the defining characteristics of our new security environment. During the Cold War, we faced a fairly predictable set of threats. We came to know a great deal about our adversary, because it was the same one for a long period. We knew many of the capabilities they possessed, and we fashioned strategies and capabilities that we believed we needed to deter them. And they were successful. It worked. For almost a half a century, that mix of strategy, forces and capabilities allowed us to keep the peace and to defend freedom. But the Cold War is over. The Soviet Union is gone, and with it, the familiar security environment to which our nation had grown accustomed. As we painfully learned on September 11th, the challenges of a new century are not nearly as predictable as they were during the Cold War. Who would have imagined only a few months ago that terrorists would take commercial airliners, turn them into missiles and use them to strike the Pentagon and the World Trade Towers, killing thousands? But it happened. And let there be no doubt, in the years ahead, it is likely that we will be surprised again by new adversaries who may also strike in unexpected ways.
>
> (Rumsfeld, 31/1/2002)

As noted above, this statement provides a useful demonstration of the continuity of this transformatory narrative more than four months beyond the attacks' occurrence. As in Wolfowitz's earlier diagnosis, Rumsfeld again invokes 9/11 as a bringer of a new security environment comprising unfamiliar and irrational enemies determined to strike the US in (equally) unprecedented ways. Although slightly more circumspect on the possibility of such a surprise (Rumsfeld's 'likely' here replacing Wolfowitz's 'virtual certainty'), this constructed discontinuity is once more presented in effectively objective phraseology: as a factual statement of probability, not an individual interpretation derived from a particularistic, retrospective, reading of 9/11's status and import. Where Wolfowitz's statement juxtaposed this new security environment to an implicit Cold War imaginary of rational stability, however, Rumsfeld's makes this juxtaposition quite explicit. With the certainties of (successful) security through deterrence now, seemingly, lost forever, a (new) post-Cold War era of unimaginable foes appeared with us to stay (see, for example, Bush, 11/10/2001b, 13/6/2002; Rumsfeld, 31/1/2002; White House, 9/2002: 29). And, as Condoleezza Rice reflected in a 8 September 2002 interview with CNN, this new security environment would undoubtedly require the readiness to confront emergent threats through whatever means possible:

> The single most important lesson that I've learned is that, unfortunately, you will always be surprised about the magnitude of events; that you will be surprised, particularly in this world, with terrorism and weapons of mass destruction, at how much damage can be done by a few people; and that you should not wait to be surprised by evil people who may wish you real harm with weapons of mass destruction that would make September 11 look small in comparison. History shows us that inaction is the problem, and the vulnerability of the United States is really what came home very, very clearly on 9/11. We've been a country that's been fortunate to be protected by two oceans, to not have serious attacks on our territory for most of our history. And we were unfortunately reminded in a very devastating way of our vulnerability. We're in a new world. We're in a world in which the possibility of terrorism, married up with technology, could make us very, very sorry that we didn't act. So I think, if September 11 taught us anything, it taught us that we're vulnerable, and vulnerable in ways that we didn't fully understand.
>
> (Rice, 8/9/2002)

The politics of warnings and lessons

As we have seen in this discussion, 9/11 was persistently positioned as a moment of abrupt temporal discontinuity by representatives located throughout the Bush administration – not only in the attacks' immediate aftermath, but some distance beyond. Signalling the end of a stable Cold War period of knowable, conventional, and rational adversaries, 9/11 was produced here as a warning of a new vulnerability marked not only by global terrorist actors but also their unpredictable, unnameable equivalents. A critical event separating an old time of predictable, manageable, security from an entirely new time within which surprise and contingency needed to be expected and guarded against. This positioning, of course, worked to extend the attacks' significance far beyond their unprecedented, tragic, impact considered above: locating this event as a truly revolutionary historical moment ushering in 'a set of circumstances unlike any we've ever faced' (Cheney, 29/11/2001).

This repeated effort to locate 9/11 as a temporal discontinuity was not, of course, an inevitable or necessary product of the attacks themselves. Most obviously, it relied on a particularly narrow, a particularly exclusionary, reading of the pre-9/11 environment as one of ordered stability and relative security. An alternative reading of that world mobilising a broader understanding of violence or security than that apparent in the above discussions would have rendered this construction of the attacks questionable, if not simply unacceptable. For, as we noted above, insecurity and violence represent an unfortunately daily reality for many individuals and communities throughout America and beyond. Even if, however, we were to accept only the rigidly narrow, rigidly realist, conception of American *national* security evident in Rumsfeld's 31 January 2002 remarks, for example, it would quite clearly be possible to challenge both the historical and analytical parameters of his argument. For, as Padelford's (1951: 388–9) discussion of American insecurity indicates rather neatly, it was certainly possible to read the Cold War period through a rather different lens:

> The United States, striving to 'contain' Soviet imperialism, feels insecure as it is confronted with the Soviet Union's larger and more tightly disciplined population, its backlog of untapped resources, and particularly its advantageous geographical position enabling the Soviet Union to thrust at will over contiguous land areas into Europe, the Middle East, and Asia. ...There was a time when the United States and other countries could feel reasonably secure in following policies of isolation. Such a situation, however, has not existed since fascist and communist states have embarked upon programs of territorial and political aggrandizement aimed at altering the entire balance of world power.

Even if we were to accept this contentious and partial reading of the pre-9/11 era as a time of relative security, it would also, clearly, be possible to question the inductive logic underpinning the projection of this complex and discrete historical event forwards as an indicator of further present/future insecurity. As noted above, alternative writings of the attacks as an anomalous, atypical, freak occurrence would clearly have challenged the prophetic qualities attributed to 9/11 here. As would the more explicitly linear constructions of those attacks as simply the latest moment in a history of ever-more dangerous unconventional violences that we turn to in the next chapter. Although, of course, impossible to falsify this inductive reasoning empirically – particularly with respect to the more ambiguous writings of this new security environment– it is important that we note the movement involved in this first isolation of 9/11 as an indicator or agent of future insecurity. For, this attempt to project forwards characteristics attributed to the 9/11 attacks (unforseeability, irrationality, and so forth) as indicative of a new present/future depended, of course, on the mobilisation of a particular reading of that event, and on a particular conception of temporality as a periodically discontinuous process. In this sense, it is important we remember that this ostensibly objective production of 9/11 as a dividing point between two horizons of security and threat must be approached as a contingent, artifactual, and therefore political, construction.

If we are willing to approach this writing of 9/11 as a political construction by virtue of the exclusionary implications it posed for alternative interpretations of those events, it is vital that we also pay attention to its constitutive functions within the unfolding War on Terror itself. Most obviously, this writing worked to legitimise some form of responsive action to the attacks themselves – not as a retaliatory gesture, but, rather, as a necessary reaction to the new security environment in which we now found ourselves. Manufacturing and selling the War on Terror as a necessary product of the attacks would quite clearly have been far more difficult on normative *and* political grounds had this event been presented as an anomalous occurrence with little chance of recurrence. For, as Colin Powell (16/9/2001) suggested in his interview on CBS:

> I assume that there are those out there who are still planning activities against the United States whether we retaliate or not. We should not see this just in terms of retaliation for the sake of retaliation, just to strike for the sake of striking. We should see it in terms of a campaign that goes after not just retaliatory satisfaction, but goes after eliminating this threat by ripping it up, by going after its finances, by going after its infrastructure, by making sure we're applying all the intelligence assets we can to finding out what they may be up to. The measure of success at the end of the day will be no more attacks likes this or over any other nature against the United States and our interests around the world.

John Ashcroft, similarly, argued that it was imperative we began preparations to avoid another 9/11 in his 6 December testimony before the Senate Committee on the Judiciary:

> Since those first terrible hours of September 11, America has faced a choice that is as stark as the images that linger of that morning. One option is to call September 11 a fluke, to believe it could never happen again, and to live in a dream world that requires us to do nothing differently. The other option is to fight back, to summon all our

> strength and all our resources and devote ourselves to better ways to identify, disrupt and dismantle terrorist networks.
>
> (Ashcroft, 6/12/2001)

As did the President, in rather more folksy language, on 4 October 2001: 'People understand that now is the time to take a stand, to seize this moment, to say that out of this evil action will come good' (Bush, 4/10/2001b). And in his subsequent, 12 November 2002, reminder of the stark choice offered by the lesson of 9/11:

> It's a new charge. It's a new charge because we learned on that fateful day that America is now a battlefield. It used to be that oceans would protect us. We didn't have to take certain threats seriously. We could say, well, we can deal if we want to deal with them. But we learned a tough lesson, that the old ways are gone, that the enemy can strike us here at home, and we all have new responsibilities.
>
> (Bush, 12/11/2002)

In short, this discontinuous writing of 9/11 as a bringer of new threat and insecurity rendered doing nothing a simply naive, irresponsible, and potentially devastating course of action for the American government. If those attacks had indeed sounded the final death knell of the Cold War era once and for all, they also 'made it clear that the US must use every tool at its disposal to combat terrorism' (Powell, 26/12/2001).

There will, of course, be many among us for whom doing nothing was indeed simply unthinkable in light of the destruction wrought by 9/11. For those, the legitimating functions offered by this writing of a new security environment may appear largely incidental: the Bush administration, it could be argued, was always going to react somehow to the deaths of Americans on American soil – such is the responsibility and mandate of government. Yet if we turn to the ways in which this discontinuous writing also worked simultaneously to justify many of the more controversial and pernicious transformations in American security practices that followed the events of that day, it becomes manifestly clear that we are not dealing here with mere rhetorical dressing. For, with the Cold War techniques of deterrence and containment seemingly obsolete in the face of this new (irrational, ruthless, and bloodthirsty) aggregation of enemies, we were made aware very quickly that treaties, agreements, and the like could no longer be relied on to keep us secure (see, for example, Cheney, 23/10/2001, 7/8/2002, 9/4/2003). With a radically new security environment upon us, only a systematic, pre-emptive, engagement with these new threats would afford America(ns) and others security. Diplomatic efforts, for example, were always already destined to fail.

This is not the place for a comprehensive account of the manifold developments that were instituted under the War on Terror rubric in the months following the September 11 attacks. There are already a number of valuable surveys performing such a function (see, for example, Scraton, 2002; Chomsky, 2003; Cole, 2003; Bellamy et al., 2008). What is here appropriate, however, is to trace the ways in which some of these measures were persistently justified by recourse to this first discontinuous writing of 9/11 as a warning and lesson. So, if we look first at domestic initiatives launched by the Department of Justice, for example, Ashcroft (25/9/2001) was already arguing for novel and expanded legislation to combat this new threat just 14 days after 9/11's occurrence: 'Every day that passes with outdated statutes and the old rules of engagement is a day that terrorists have a competitive advantage. Until Congress makes these changes, we are fighting an unnecessarily uphill battle.' Little wonder, then, that he returned to this theme on the eve of the signing of the now-infamous Uniting and Strengthening America by Providing Appropriate Tools Required to Intercept and Obstruct Terrorism Act of 2001:

> The president is expected to sign this legislation on Friday. The hour that it becomes law, I will issue guidance to each of our 94 U.S. Attorney's Offices and 56 FBI field offices directing them to begin immediately implementing this sweeping legislation. I will issue directives requiring law enforcement to make use of new powers in intelligence gathering, criminal procedure and immigration violations. A new era in America's fight against terrorism, made tragically necessary by the attacks of September 11, is about to begin.
>
> (Ashcroft, 25/10/2001)

And, again, in his later, 9 September 2003, reflections on what had become a period of quite dramatic legislative transformation within the United States:

> September 11 taught us that terrorists had outflanked law enforcement in technology, communications and information. So we fought for the tools necessary to protect the lives and liberty of the American people. Congress provided these tools in the USA Patriot Act, passed by overwhelming, bipartisan majorities ... 98 to 1 in the

> United States Senate and better than a 5 to 1 margin in the United States House of Representatives. And while our job is not finished, we have used the tools provided in the Patriot Act to fulfil our first responsibility to protect the American people. We have used these tools to prevent terrorists from unleashing more death and destruction on our soil. We have used these tools to save innocent American lives. We have used these tools to provide the security that ensures liberty.
>
> (Ashcroft, 9/9/2003)

Yet the impact of this writing in reshaping American counterterrorist strategy was not, by any means, limited to the Department of Justice. From the Department of Defense, similarly, this new security environment was presented as evidence of the need for a comprehensive transformation of the American military and defence strategising (Croft, 2006: 136–8). Donald Rumsfeld's (31/1/2002) 31 January 2002 remarks at the National Defense University, for example, presented 9/11 as vindication of the '6 transformational goals' he had earlier established 'to keep the peace and defend freedom in the 21st century':

- First, to protect the U.S. homeland and our bases overseas.
- Second, to project and sustain power in distant theaters.
- Third, to deny our enemies sanctuary, making sure they know that no corner of the world is remote enough, no mountain high enough, no cave or bunker deep enough, no SUV fast enough to protect them from our reach.
- Fourth, to protect our information networks from attack.
- Fifth, to use information technology to link up different kinds of U.S. forces so that they can in fact fight jointly.
- And sixth, to maintain unhindered access to space and protect our space capabilities from enemy attack.

With his subsequent, 6 August 2002, remarks providing a more explicit account of this strategic and tactical transformation that had been precipitated by this abrupt end to the certainties of the Cold War period:

- We adopted a new approach to strategic deterrence that increases security while reducing our dependence on strategic nuclear weapons. The missile defense research and testing program has been reorganized and revitalized, free of the constraints of the ABM Treaty.

- We're investing in a range of new capabilities that should help us better defend our country in a dangerous and uncertain period that's ahead.
- We also fashioned a new Unified Command Plan and reorganized our worldwide military command structure and strengthened our focus on homeland security.

(Rumsfeld, 6/8/2002)

In fact, whichever example we choose to consider: be it pre-emptive military action (Bush, 14/6/2002; Rice, 1/10/2002; Rumsfeld, 4/2/2002; Wolfowitz, 1/6/2002), the establishment of the Office/Department of Homeland Security (Bush, 18/6/2002; Ridge, 26/9/2002), the detention and treatment of suspected terrorists (Bush, 29/11/2001; Cheney, 29/11/2002), or the emergence of new and increased surveillance powers (Bush, 16/6/2006, 7/9/2006), appeals to the lessons of this new security environment were never far away. Claims to temporal discontinuity and newness, it seems, offer a remarkably useful discursive resource with which to push through transformations in regimes of governance.

If this writing of 9/11 worked to legitimise the War on Terror as a totality, and to justify many of its more concrete, controversial, developments besides, these political functions did not exhaust its utility for the Bush administration. On top of the above, this scripting of a new world characterised by unidentifiable, unnameable, security challenges opened substantial discursive space for expanding this conflict beyond its initial targets of terrorists and their abetters. Positioning 9/11 not only as an indicator of future terrorist attacks but also as a vague, ill-specified, warning of a generalised present/future insecurity, this discontinuous writing was absolutely integral to the subsequent incorporation of rogue states and their equivalents into the parameters of this conflict. Most obviously, most damagingly, it laid the foundation for the commencement of military activities against the Iraqi regime in March of 2003: a commencement of military activities that was frequently predicated on the claim that 'The expectation of the familiar must not guide us as we move forward. Rather the unfamiliar and the unlikely must be our new guides' (Wolfowitz, 19/9/2002). And, as the President explicitly noted in his 2003 State of the Union Address:

> Before September the 11th, many in the world believed that Saddam Hussein could be contained. But chemical agents, lethal viruses and shadowy terrorist networks are not easily contained. Imagine those 19 hijackers with other weapons and other plans – this time armed

> by Saddam Hussein. It would take one vial, one canister, one crate slipped into this country to bring a day of horror like none we have ever known. We will do everything in our power to make sure that that day never comes.
>
> (28/1/2003)

Finally, this particular framing of temporality as a discontinuous process also offered a powerful contribution to the processes of collective identity construction already underway in the above-discussed constructions of rupture. By positioning 9/11 as a bringer of new threat and insecurity not only to Americans collectively but also to civilisation and the world more broadly, this writing both drew on and reinforced the antagonistic, dichotomous, division between us and them upon which the unfolding War on Terror depended. Not only had 'we' (Americans, freedom lovers, civilisation, the world) collectively experienced a moment of unspeakable, unpredictable, tragedy and evil on 11 September 2001. 'We' were also now equivalent in our subjection to the new security challenges of this confrontation that extended far, far beyond those initial attacks and their immediate protagonists. Questioning or challenging the subsequent responses to 9/11 was not only counterproductive in such a climate but also a potentially suspicious, unpatriotic denial of this shared identity: as Bush (20/9/2001, 4/10/2001b, 6/11/2001, 5/1/2002) repeatedly argued: 'You are either with us, or you are with the terrorists.' And, for those advocates of public dissent and resistance within this new security environment, John Ashcroft (1/10/2002) offered the following warning:

> Our critics seem to think that business-as-usual – doing what was done before – and nothing more – would keep America safe from terrorists. As my grandfather used to say, 'I've sawed this board off four times and it's still too short.' If we keep on doing the same thing, we cannot be surprised by the same results. History instructs us that caution and complacency are not defenses of freedom: caution and complacency are a capitulation before freedom's enemies – the terrorists. While we always assiduously respect civil liberties and rights; do not become timid or shrink from your duties because of slings and arrows in the public arena.

As we have seen, the Bush administration moved with remarkable rapidity to position 9/11 as a moment of temporal discontinuity in the attacks' aftermath. Expanding on earlier claims to their unprecedented, rupturing, destabilising impact, this writing located those events as a coherent and identifiable temporal interval separating an old time of predictable, manageable security from a new (present/future) time of radical insecurity. A new time characterised not only by irrational, undeterrable terrorist groups and their sponsors but also by weapons of mass destruction, rogue regimes, and any number of additional unforeseeable threats. That this writing was both contestable – a contingent formulation, rather than an inevitable product of those attacks – and, indeed, productive – contributing to the legitimisation, expansion, and identity politics of this war – offers further evidence for the linkages between time, violence, and the political touched on already above. How we imagine temporality – whoever we are – poses considerable implications for the social worlds that we produce. And, as we shall now see, this writing of 9/11 as warning/lesson was far from unique in the administration's efforts to construct their new, unfolding war around claims to temporality as a discontinuous, punctuated process.

Writing 9/11 as declaration/instigation of war

On 13 September 2001, an unidentified speaker asked President Bush about his emergent response to the attacks of two days prior. Questioned, specifically, on his attempt to forge an international coalition akin to that constructed by his father in what is now known as the first Gulf War, the President's response included the following, prescient, remarks:

> I've been on the phone this morning, just like I was yesterday, and will be on this afternoon, on the phone with leaders from around the world who express their solidarity with this nation's intention to rout out and to whip terrorism. They understand, fully understand that an act of war was declared on the United States of America. They understand, as well, that that act could have as easily been declared on them; that these people can't stand freedom; they hate our values; they hate what America stands for. Many of the leaders understand it could have easily have happened to them ... there is universal support for the American people, sadness in their voice; but understanding that we have just seen the [first] war of the 21st century.
>
> (Bush, Pataki & Giuliani, 13/9/2001)

In several key ways, Bush's remarks here reinforce everything we have already learned about 9/11. Once again, those attacks are positioned as an unjustifiable assault on America (and American values). Once again, we are confronted with the claim that the import of that assault extends far beyond the borders of the United States alone. And, once again also, we are left with the expectation of an imminent, and aggressive, response to those attacks. In these respects, this passage clearly strengthens the sense of temporal discontinuity we have been exploring throughout this chapter. Something significant, it seems, changed on that day: something new was being brought into existence amidst the trauma, destruction, and wreckage we were still struggling to name and comprehend.

There is, however, a crucial difference to note in comparing Bush's reflections here with those we have already considered. Not only is the President locating 9/11 as a bringer of new threat and insecurity in these remarks, he is also locating that event as both a declaration and instigation of war itself. Positioned in this way as the instituting, originary moment of the 'first war of the 21st century' (see also, Bush, 17/9/2001a; Bush & Megawati, 19/9/2001; Bush, 24/10/2001), the attacks here mark our entry into a new time not only of danger but also of conflict. Abstracted from any historical contextualisation (of ongoing conflict), and, indeed, from any specific, direct, political justification (war could have been as easily declared on any of the other states contacted by Bush), 9/11 now emerges as a military assault that simply could not but have catapulted the United States (and its allies) into conflict with terrorism. War, it appears, arrived on American soil that day, unexpectedly, without warning. War had already begun on that morning, irrespective of American intentions and actions. Yet if this absence of any explicit warning or motive leaves any doubt as to their normative reprehensibility, Bush prefaces the forthcoming 'whipping' of terrorism by relaying the patient processes of dialogue and consensus building in which he had already been engaging. Even in light of such a tragic, evil attack, then, the republican administration had responded in a controlled, reflective, and even restrained manner.

This rather powerful construction of the 9/11 attacks as a declaration and instigation of war on America was returned to with profligacy by the President throughout the weeks and months that followed (see, for example, Bush, 14/9/2001, 17/9/2001a, 25/9/2001, 1/10/2001, 17/5/2002). Speaking at Camp David on 15 September, for example, he returned to this positioning, in reflecting on his recent visit to New York:

> I am going to describe to our leadership what I saw: the wreckage of New York City, the signs of the first battle of war. We're going to meet and deliberate and discuss – but there's no question about it, this act

> will not stand; we will find those who did it; we will smoke them out of their holes; we will get them running and we'll bring them to justice.
>
> (Bush, Powell & Ashcroft, 15/9/2001)

As he did in his subsequent remarks at the Pentagon on the attacks' first anniversary, albeit with considerably greater emotive force:

> What happened to our nation on a September day set in motion the first great struggle of a new century. The enemies who struck us are determined and they are resourceful. They will not be stopped by a sense of decency or a hint of conscience – but they will be stopped. ... In every turn of this war, we will always remember how it began, and who fell first – the thousands who went to work, boarded a plane, or reported to their posts.
>
> (Bush, 11/9/2002a)

The Vice President, similarly, appeared to conceive those attacks as marking the emergence of America's involvement in a new conflict, referring repeatedly to 'the war that began on September 11' (Cheney, 6/6/2002, 19/7/2002, 26/8/2002, 2/10/200, 17/9/2003), and arguing, 'We did not seek this war, but it found us prepared' (Cheney, 10/11/2001, 18/2/2002). As, too, did those working at the Pentagon in the Department of Defense (see, for example, Clarke, 28/10/2001), with Secretary Rumsfeld (1/11/2001) not only warning America(ns) to prepare for what was likely to be a long-running confrontation:

> With the ruins still smoldering and the smoke not yet cleared, it seems to me that Americans understand well that – despite the urgency in the press questions – we are still in the very, very early stages of this war. The ruins are still smoking!

But, also, reminding them of the stakes for which this conflict was being conducted:

> The terrorists who attacked the United States on September 11 did more than blow-up buildings, or murder thousands of innocent people. They declared war indeed on our way of life and they attacked us because I suppose of what we are. As Americans we are free people, free men and women, proud of our country's cause and the cause of human freedom. In so doing, they reminded us that the world is

> a dangerous place, it's an untidy place, it's a difficult place and it's a changing place where determined adversaries still oppose human freedom and will stop at literally nothing to destroy it and deny it. And so long as such enemies exist freedom everywhere will be in danger.
>
> (Rumsfeld, 9/6/2002)

Over at the Department of Justice, John Ashcroft (18/9/2001), likewise, positioned the attacks as 'nothing short of a declaration of war against the people of America', in a departmental briefing just one week after the attacks. While Tom Ridge, the new representative of the administration's Homeland Security project, argued similarly on 30 October: 'we're all very much aware that on September 11th our war against terrorism began' (Ridge et al., 30/10/2001). Indeed, throughout the immediate aftermath of the attacks and for some time beyond, only Colin Powell (12/9/2001c) appeared keen to offer a more equivocal reading of their historical location, suggesting on the day after their occurrence:

> It doesn't mean necessarily that it results in, say, a declaration of war – Congress would have to make that judgment anyway – but it means that we have to really mobilize ourselves and all of the assets at our disposal – political, diplomatic, legal, law enforcement, intelligence and military – to deal with those who perpetrated this act of war against us.

Powell's comments are particularly interesting because of the challenge they offer to an otherwise coherent and pervasive writing of 9/11. As we have seen, the Bush administration worked hard to position those events as a self-evident temporal interval separating an old time of non-war from a new time of war. For the moment, however, I would like to leave this challenge hanging, and turn to an important sense in which this discontinuous writing was further reinforced by these same political actors speaking at the same time. For, alongside the above claims to war's commencement that day, we were also repeatedly told that this new war would be emphatically, qualitatively, distinct from any we had previously seen. This representation, already evident in Bush's 13 September telephone conversation with which we began this discussion, further vindicated these claims to our recent witnessing of a world-historical shift. It did so, by presenting 9/11 now as a moment of transition that had launched America and its allies not only into a new (unsought, unexpected) conflict but also into an entirely unprecedented form of conflict (see McInnes, 2003).

On 17 September 2001, President Bush visited the Pentagon to address employees at one of the four sites that had been attacked on 11 September. His remarks on that occasion offered his then fullest account of the unfolding War on Terror's likely conduct and progress, drawing an explicit distinction in the process with earlier confrontations to which the United States had been party:

> I believe, I know that an act of war was declared against America. But this will be a different type of war than we're used to. This is – in the past there have been beaches to storm and islands to conquer. We've been able to watch on our television screens sophisticated weaponry find a building; and we've seen dramatic reports from the front where Pulitzer Prize-to-be winning reporters stood up and declared, the United States is attacked, and all that. There may be some of that, who knows. But I know that this is a different type of enemy than we're used to. It's an enemy that likes to hide and burrow in, and their network is extensive. ... But we're going to smoke them out. And we're adjusting our thinking to the new type of enemy. These are terrorists who have no borders.
>
> (Bush, 17/9/2001a)

As in many of his reflections of this time and later, Bush here positions the War on Terror as an unprecedented form of conflict (Bush, 25/9/2001, 4/10/2001a, 11/6/2002, 10/7/2002). Contrasting this emergent conflict, explicitly, with earlier battles, the President lets us know that we are witnessing something of a revolution in the conduct and conditions of warfare. Confronted by a new (stateless, amorphous, barbaric) aggregate of enemies, the capture of territory could not, and would not, be seen as indicative of success (Bush, 15/9/2001, 16/9/2001, 17/9/2001a; Rumsfeld et al., 27/9/2001). With conventional military superiority no longer a guarantor of victory (Cheney, 16/9/2001), a conflict of often-unseen battles and progress was required (Bush & Megawati, 19/9/2001; Bush, 20/9/2001, 26/9/2001b, 29/9/2001). A comprehensive war of multiple, fluid fronts (Bush et al., 24/9/2001; Bush, 2/11/2001, 12/11/2002) characterised by unprecedented national and international cooperation (Bush, 12/9/2001c, 10/10/2001b; Powell, 10/10/2001, 22/5/2002, 7/12/2002) and lacking any discernible, knowable endpoint (Bush, 11/10/2001b, 1/5/2002; Cheney, 16/9/2001). 9/11, it seemed, had not only catapulted the United States and its allies into a new, unforeseen war, it had also catapulted them into a qualitatively new type of conflict. In this sense, it is possible now to identify the clear and explicit discursive linkage between the writing of 9/11 as a bringer of new insecurity considered above and this related writing of 9/11 as a bringer of new war: 'A new kind of war against a new kind of enemy' (Cheney, 24/5/2002), a 'war unlike any other war we have fought' (Ashcroft, 19/2/2002). For, as Rumsfeld (12/9/2002) suggested in his 12 September 2002 remarks at the Funeral Service for unidentified victims of the Pentagon attacks:

> Those we honor today died here at home – not on a faraway battlefield. They died within view of this cemetery. Yet they did die on a battlefield – and that battlefield tells us a great deal about the war we are in – the first war of the 21st Century.

The politics of war declarations

We approached the first writing of 9/11 as a warning of new insecurity as a distinctly political construction in both an exclusionary and constitutive sense. As outlined there, the persistent positioning of the attacks as a particular moment of temporal discontinuity worked to militate against alternative conceptions of those events, while also rendering the War on Terror both possible and legitimate. In similar vein, this second writing of those events as a bringer of a (related) radically new time of (new) war against terrorism must also be viewed as inherently political: despite, of course, the claims to objectivity around which this framing was frequently couched.

Most obviously, this writing of 9/11 as the War on Terror's originary moment worked once again to isolate those attacks from any earlier, and potentially significant, occurrences or historical trends. As with the first writing of new insecurity, this positioning again built on the continuing effort to depict 9/11 as a moment of temporal rupture: projecting the attacks' singular significance forwards as a(nother) harbinger of radical, qualitative change. So exceptional were those attacks, in this presentation, that they were simply external to anything we had witnessed before. As something unique had occurred on that day, something new had also begun. Yet, as Feith and Bush both subsequently, explicitly, indicated, it would indeed have been quite possible to locate 9/11 otherwise by pointing to earlier indicators of this ostensibly new confrontation:

> President Bush's statements even then showed that he thought of the attack, in essence, as an act of war, rather than a law enforcement matter.

> That point may now seem unremarkable, but think back to the 1993 World Trade Center bombing and to the attacks on Khobar Towers in 1996, on the US East Africa embassies in 1998 and on the USS Cole in Yemen in 2000. When such attacks occurred over the last decades, US officials avoided the term 'war'. The primary response was to dispatch the FBI to identify individuals for prosecution. Recognizing the September 11 attack as war was a departure from established practice. It was President Bush's seminal insight, the wisdom of which is attested by the fact that it looks so obvious in retrospect.
>
> (Feith, 13/11/2003)

> I know that some people question if America is really in a war at all. They view terrorism more as a crime, a problem to be solved mainly with law enforcement and indictments. After the World Trade Center was first attacked in 1993, some of the guilty were indicted and tried and convicted, and sent to prison. But the matter was not settled. The terrorists were still training and plotting in other nations, and drawing up more ambitious plans. After the chaos and carnage of September the 11th, it is not enough to serve our enemies with legal papers. The terrorists and their supporters declared war on the United States, and war is what they got.
>
> (Bush, 20/1/2004)

As these more recent reflections (and, indeed, Powell's earlier remarks noted above) quite vividly suggest, this writing of 9/11 as a bringer of new war was neither an inevitable nor referential description of those events. Rather, this discontinuous framing of those events was a (contingent, violent, and contestable) decision taken against an undecidable ontological horizon. Alternative readings of those events – of that horizon – emphasising their continuity with earlier attacks would, quite clearly, have posed considerable consequences for the shape and conduct of the War on Terror itself. For, although legal papers, prosecutions, and FBI operations may not have appealed to the Bush administration at the time, the overlooking of those alternatives was a product of will: emphatically not a product of necessity.

Bringing into view the moment of decision involved in this (partial, particular) writing of temporality becomes even more crucial when we consider a second exclusion upon which it depended. For, by framing the events of 9/11 as a bringer of (new) war to the American public, this writing also militated against any recognition of the Bush administration's agency in launching the War on Terror.

Depicted thus, the Bush administration not only responded correctly or appropriately in confronting this new enemy. They simply had no alternative: war had *already arrived* at the United States' borders that morning. In an important sense, then, political dissent or resistance to this emergent security paradigm was once more already being challenged before it had taken root. If we were not yet entirely sure what it would look like, we could be sure that the War on Terror had already begun. And yet, interestingly, as we shall see in the following chapter, this stripping of any meaningful agency from the Bush administration's decision to engage in this conflict was followed by a persistent and repeated attempt to reassert a sense of control over this unfolding war with the writing of the War on Terror once underway as a story of continuing, linear, progress led by the United States' government.

If this writing of temporal discontinuity depended on the narrative exclusion of alternative pasts, presents, and futures relating to the events of 9/11, it was also, again, remarkably productive for the Bush administration's conduct of the War on Terror itself. Most obviously, this positioning worked once more to (re)inscribe significance into the 9/11 attacks: positioning this event as important by reaffirming the sense of disjunction noted throughout this chapter thus far. Where the writing of new insecurity achieved this by stressing the attacks' impact in catapulting the United States (and their allies) into an unprecedented security environment, this writing presented 9/11 as significant because of its status as a bringer of (a fundamentally new type of) conflict. It is hard to think of any more significant historical events than those moments in which wars begin or are launched. And, by conferring this identity on those four crashes, Bush's administration here further reinforced the sense that we had witnessed a genuinely world-defining historical moment that day.

Perhaps even more important for subsequent events, however, was the way in which this writing also worked to structure the form of response that the War on Terror would subsequently comprise. By problematising the 9/11 attacks within a specific, militarised, frame – as the first (illegal, unjustifiable) act of a new kind of conflict – this writing served as a pre-emptive justification for the subsequent launch of military actions by the United States: initially in Afghanistan but rapidly further afield. Rendering any form of legal or diplomatic proceedings already redundant in the aftermath of the onset of this conflict, the Bush administration appeared to have little choice but to resort to their military power if defeat was to be avoided. As war had begun on

11 September 2001, war had to continue – unfortunate and unpopular as that may have been at home and abroad. If we take this in conjunction with the bringer of new insecurity frame considered above, then, this multifaceted positioning of 9/11 as a moment of temporal discontinuity not only rendered War on Terror necessary and legitimate. It also, crucially, helped shape the parameters of that conflict from its very emergence.

Third, although this construction of 9/11 as a declaration or institution of war worked to problematise that event within an already productive frame of military conflict, it also achieved far more even than this. By reminding us, repeatedly, of the emphatically new character of the war that arrived on 11 September 2001, this writing rendered the entire panoply of military, financial, diplomatic, and legislative actions that subsequently characterised the War on Terror discursively equivalent. Each of these developments emerged here as contributory moments within the same broad, unprecedented, historical problematic. In both reaffirming the significance of military measures for this conflict and, paradoxically, drawing our attention to their limited utility, then, this production of temporal discontinuity also offered considerable structural coherence for this new type of war.

Finally, and perhaps most important of all, this production of 9/11 also functioned to inscribe *normative* legitimacy into the subsequent War on Terror itself. By positioning the attacks as the first (unforeseeable, unjustifiable) strike in a radically new confrontation, any discussion of subsequent injustices, victimhoods, and suffering would necessarily occur in relation to this perceived act of original violence. If resorting to conflict now requires considerable justification to a diversity of audiences (Jackson, 2005: 121–2), this second construction of temporal change offered a particularly valuable mechanism for achieving this goal. To return to Bush's above remarks at the Pentagon (11/9/2002a), this writing left little room for questioning guilt and innocence in this conflict's unfolding, where, 'In every turn of this war, we will always remember how it began, and who fell first.' If, or that, further casualties would follow in this war was, perhaps, unfortunate. Yet, the responsibility for those causalities – the 'collateral damage' of this conflict – did not reside – could not reside – with Bush's administration. For that administration, and the American public, had been already attacked – already plunged into conflict – without justification or prior warning that this war was about to begin.

In sum, as with the above writing of 9/11 as a bringer of new insecurity, this second construction of radical discontinuity once again depended on a specific, contingent, construction of 9/11 as a temporal interval distinguishing an old time (of peace) from a new time (of new war). In not only excluding alternative conceptions of that event and its aftermath but also conferring structural coherence and normative legitimacy upon the subsequent War on Terror, this writing again offered an incredibly productive discursive resource for the Bush administration. In this sense, as above, it is crucial that we approach this writing of time as an inherently political construction: a violent and contestable representation that could have been plausibly and legitimately produced otherwise.

Writing self-transformation

In the chapter that follows this, we will encounter three very different governmental writings of the movement of time to those we have been discussing thus far. Before turning to those accounts of linear temporality, however, there is one further construction of temporal discontinuity we must reflect on now finally. This writing employed an identical temporal structuration to those of new insecurity and new war we have been exploring here: producing 9/11 again as an interval between an old time and a new. But, it also differed, interestingly, from either of the above in locating those attacks' transformatory impact exclusively *within* the borders of the United States. For, as we shall see below, the Bush administration also attempted to locate 9/11 as the end of an era of selfish individualism, material self-interest, and other undesirable social characteristics within the American population. And, at the same time, to locate the attacks as marking the start of an entirely new period of harmony, unity, and collective identity.

Beyond its resonance with the representations considered thus far, there are three further reasons for my interest in this third invocation of temporal discontinuity by the Bush administration in their writing of the War on Terror. In the first instance, where each of those considered above invoked – or made possible – a subject of change extending beyond the United States' borders, this writing was exclusively structured at the level of the domestic or national. If America and its allies had been plunged into a new era of perpetual insecurity, if America and its allies had been plunged into a new (type of) war, America alone, it seemed, had been internally changed by those attacks. If the significance of 9/11 stretched across the globe, some dimensions of the attacks' impact, at least, appeared rather more contained.

Second, where each of the above writings presented the discontinuity effected by 9/11 as an inherently and entirely negative phenomenon, this particular construction also introduced something of a normative ambiguity into those events. Positing beneficial consequences alongside the insecurity and war ushered in by this tragedy, these claims to self-transformation worked to identify a glimmer of hope residing within even the immediacy of this crisis. 9/11, it seemed, was not fully or completely harmful to America(ns): something positive appeared to also have followed in the wake of those attacks. And, finally, and perhaps most interesting of all, this third writing of temporal discontinuity was far less prominent than either of those considered above. Although Rumsfeld (11/9/2002), for example, was able to argue of 9/11, 'there's no question but that it did awaken a great deal of patriotism and respect for what we have and how valuable it is, our liberty and our freedom', President Bush and Tom Ridge were almost unique within the administration in seeking to position those events as a bringer of national, *desirable*, change. Whether the relatively restricted usage of this writing speaks to its limited resonance with the audiences to which it was targeted. Or, perhaps, to its lesser productivity for the War on Terror itself, it will be worth keeping this key distinction in mind. For, as outlined in the following chapter, the administration was far more keen to employ an entirely different conception of post-9/11 America(ns) in the aftermath of those events: a conception structured around an incompatible conception of temporality to that which we are here exploring.

On 11 October 2001, the President launched the *America's Fund for Afghan Children* as part of his administration's unfolding engagement within Afghanistan. In a discussion of that new programme that took place the following day, Bush (11/10/2001c) justified its existence with a brief revisiting of earlier troubling times encountered by America(ns). Recounting, here, the experience of the Great Depression upon American people, the President pointed to an identifiable national spirit that seemed to have been lost in the intervening years. But, a spirit, intriguingly, that now looked as though it were being regained:

> At the height of the Depression, one-third of Americans were unemployed. Nearly 40 percent of banks had failed. And there was great anxiety and uncertainty in our country. Yet, hard times drew people together. The Great Depression tested America's character and revealed America at its best. Americans have shown a similar strength since September the 11th. Terrorists hoped our nation would come apart. That's what they hoped for. But, instead, we've come together. Our country is more resolved, more united and guided by a greater sense of purpose than any time during our lifetimes. And some

> important things about our culture seem to be shifting. After the attacks, moms and dads held their children closer. And maybe for a moment longer. Millions have gone to synagogues and churches and mosques to renew their faith, to find perspective, to be reminded of the true values of life.

Bush's remarks here explicitly position 9/11 as an agent of self-transformation on two distinct yet related grounds. In the first instance, he presents the attacks as having fostered a (relatively) unprecedented spirit of unity and identity within the American people: the American people guided now by a newfound and tangible sense of togetherness, purpose, and resolve. As hard times in the past had constituted a sense of community or union within that great country, so the recent tragic experiences seemed to be doing something similar. In spite of the efforts of those seeking (and here failing) to sow disharmony through their evil actions, then, the precise opposite had happened, and a new, united, America had emerged (see, for example, Stoddard & Cornwell, 2002; Silberstein, 2002: 12–16; Jackson, 2005: 85–8).

Bush's remarks also identify a second transformatory impact of the attacks in his locating a related 'cultural shift' within the American population confronted by these recent 'hard times' (see also, Bush, 8/10/2001, 4/12/2001). This (re)discovery of 'the true values of life', including faith, an appreciation of the family, and a sense of 'perspective', stretches here across the diversity of America's (religious) population: Christians, Jews, and Muslims together united in their search for meaning in the wake of such a 'trauma'. As with the above invocation of social 'unity', however, this depicted cultural shift again emerges through the positing of an oppositional, antagonistic, relationship: setting the American public's honest and legitimate quest for family and faith against the evil pervertors of Islam that had manipulated religion as a tool for their own instrumental purposes. It is this writing of piety and goodness into the American public, then, that here underpins Bush's instituting the *America's Fund for Afghan Children* as a reminder of American generosity even at this time of (reluctant, unwanted, and undeserved) conflict.

As noted above, Bush returned to this writing of 9/11 as self-transformation a number of times throughout the following months. Whether framed in terms of a new, fraternal, neighbourly kindness traversing individuals and communities throughout America – 'we must unleash the armies of compassion in every city in America to provide hope for people where hope doesn't exist ... people, as you know better than me, have said, we better assess our values as a result of what went on, and people all across the country are doing just that' (Bush, 24/1/2002; see also, Bush, 11/2/2002, 24/6/2002, 5/9/2002). Or, by reference to a new national spirit of self-sacrifice and courage:

> [W]hen we knew it was a terrorist attack – many felt that our lives would never be the same. What we couldn't be sure of then – and what the terrorists never expected – was that America would emerge stronger, with a renewed spirit of pride and patriotism. ... We are a different country than we were on September the 10th – sadder and less innocent; stronger and more united; and in the face of ongoing threats, determined and courageous.
>
> (Bush, 8/11/2001)

The President was keen to locate those events as a bringer of considerable beneficial change within America(ns). In its more poetic incarnations, indeed, this new collective spirit of responsibility was explicitly contrasted to an earlier time of hedonistic, epicurean, individualism. As the President argued in his 2002 State of the Union address, for example:

> For too long our culture has said, 'if it feels good, do it.' Now America is embracing a new ethic and a new creed – 'lets roll'. In the sacrifice of soldiers, the fierce brotherhood of firefighters and the bravery and generosity of ordinary citizens we have glimpsed what a new culture of responsibility could look like.
>
> (Bush, 29/1/2002)

A theme he returned to at Governor Pataki's reception the following month, and again in April that year:

> I know our country is beginning to make different kinds of choices. The old culture said, if it feels good, just go ahead and do it. Now we've learned a new ethic, and it had to do with Flight 93. When people on that airplane said a prayer, told their lives they loved them, and drove the plane in the ground to save others' lives, we're beginning to get a sense of sacrifice.
>
> (Bush, 6/2/2002)

> There is a new culture evolving in America. It's a culture that is shifting away from 'if it feels good do it,' and 'if you've got a problem,

> blame somebody else,' to a culture in which each of us understand that we are responsible for something greater than ourself. That was clearly defined on an airplane that crashed in Pennsylvania, when several people on that plane told their loved ones goodbye, they told them they loved them, they said a prayer, they said, let's roll, and drove an airplane into the ground to save other people's lives. It is a symbol of what is possible in America. It is a sign that there are people in this nation willing to make a sacrifice to help a neighbor in need, willing to make a sacrifice to make their community a better place.
>
> (Bush, 3/4/2002)

But, whether explicit or implicit, in each of these examples 9/11 emerges once again as a temporal dividing point separating (here) an old time of individualism and self-interest from a new time of collective unity, responsibility, and compassion.

As suggested at the start of this section, the other administration representative that appeared keen to propagate this third discontinuous writing of 9/11's impact and import was Tom Ridge from the Office and then Department of Homeland Security. His 9 November 2001 remarks at a homeland security briefing, for example, noted an emphatically new sense of civic responsibility that had been (re)awakened within the post-9/11 American public:

> I think last night, the President very appropriately tapped in to the extraordinary desire on the part of America to commit some part of themselves, in some small way, to enhance homeland security. As we have been doing every day since I took over this position, we're looking for ways to engage both the public and the private sector to help build this extraordinary support, and build toward a national strategy to provide homeland security. ... [I] think over the past several years, if you ask governors and you ask volunteer organizations, the desire, the willingness of America to participate had slowly eroded. People weren't volunteering, weren't as involved, as engaged in their community as they have been engaged before. But I think as of September 11th, there is a renewed spirit of service, a renewed spirit of public service, a renewed willingness to become engaged in supporting and providing security for your community and your country. So I look forward to the tasking of that challenge.
>
> (Ridge & Whitman, 9/11/2001)

While his discussion with newspaper editors the following April, finally, pointed to the role that his own department could play – was playing – in bringing this new sense of citizenship into being across America:

> My fondest hope is that homeland security can help spark a renewed sense of national purpose and citizenship. We no longer have the luxury of assuming that the security of our nation is somebody else's problem. More than 28,000 Americans have signed up for the President's new Citizen Corps program. They will be making a direct contribution to homeland security. Perhaps they will also be inspired to attend a PTA meeting or to mentor a young student or to vote – or, at the very least, to pick up a newspaper and learn about the outside world. In the 1950s, Americans dug fallout shelters and ducked under their desks. Today, Americans are emerging in their communities, fighting evil with acts of compassion.
>
> (Ridge, 11/4/2002)

The politics of self-transformation

As we have seen, this writing of 9/11 as a bringer of self-transformation within the United States employed the same temporal structure as the two discussed already above. Positing an imaginary, homogeneous, pre-9/11 era occupied by a singular American populace caught up in the misguided trappings of a hedonistic, materialistic, and shallow existence, the events of 11 September 2001 emerge here as a bringer of an entirely new – yet equally singular – American populace. A moment of abrupt, permanent rupture, then, in this account, ushering in a qualitatively distinct time of self-sacrifice, piety, unity, family values, and civic responsibility.

This constructed binary opposition, of course, offered an incredibly simplistic account of the American public on numerous levels. In the first instance, not only did it write out all of those individuals engaging in volunteerism and (honest, genuine) worship prior to the attacks' occurrence. It, more importantly, also worked to exclude any discussion of the (often violent) moments of public disfranchisement and disharmony that either continued or emerged after this event. Most obviously, this writing offered no location within 'the (new) American people' for those minority populations for whom life had potentially or actually deteriorated in the post-9/11 period. As Aretxaga (2002: 147–8) argued in an early reflection on these assertions of unity and national identity: 'While the meaning of this patriotic fervor is not clear and its direction is still volatile, there is little question that, for the moment, it has erased from view the scabrous issue of ongoing class and racial violence.' And, as is well known, a whole host of prominent American evangelists rushed publicly to condemn various sexual and political 'deviants' for their responsibility in helping bring the attacks about. As Jerry Falwell (cited in Croft, 2006: 71; see also Edwards, 2004: 164; Lee & Barton, 2004: 22–4) famously argued on 14 September – before subsequently apologising for his remarks:

> I really believe that the pagans, and the abortionists, and the feminists, and the gays and the lesbians who are actively trying to make that an alternative lifestyle, the ACLU, People For the American Way, all of them who have tried to secularize America. I point the finger in their face and say 'you helped this happen'.

With racist attacks and widespread ethnic profiling (see, for example, Williams, 2002: 336–7) two particularly visible consequences of the 9/11 attacks within American society, this narrative of a newly united and harmonious national community quite clearly offered a rather narrow, rather selective, writing of the American population (see, for example, Falk, 2002: 327; Sassen, 2002: 322; Huddy et al., 2003: 269; Migration Policy Institute, 2003; Ghandi, 2005: 32–3). Despite the administration's persistent and explicit condemnation of religious intolerance throughout 9/11's aftermath, then, this final discontinuous construction offered very limited space within this 'brave' new America for those individuals on the margins of American society.

If this third narrative of temporal discontinuity operated through excluding specific individuals and behaviours from the new American people, it also performed at least two key politico-discursive functions for the emergent War on Terror itself. In the first instance, by positioning 9/11 as a bearer of unintended and contingent positive implications for the American public, this construction presented the attacks as a (partial or total) failure. The (constructed) hope of the assault's perpetrators that the United States would simply 'fall apart' in the wake of this tragedy had been emphatically and demonstrably misguided according to this writing: as Bush repeatedly argued in the attack's wake: 'they misjudged us' (see, for example, Bush et al., 24/9/2001). If the most devastating and evil of attacks ever unleashed on the United States of America (according to these discontinuous writings) had proved ultimately futile in destroying this people's spirit, then, that people could remain confident of even better future times. In this sense, this writing of failure opened discursive space for the linear writings of progress into the War on Terror considered in the following chapter.

The second associated implication of this writing related, of course, to its interpellative function as a call for Americans to recognise their shared identity through engaging in this war. Reinscribing agency into a (homogenised) people that had been unwitting victims of this unforeseeable assault, and offering redemption from that people's previous excesses, this account presented a powerful and appealing pathway to a new form of (civic, social, political) righteousness that could be achieved simply through active participation in the Bush administration's War on Terror. Of particular benefit in the establishment and launch of the numerous voluntary strategies and policies that emerged at the beginning of this war's structuration, this writing offered a further useful discursive mechanism for encouraging sustained public participation in, and support for, this conflict.

In sum, although less prominent than those considered above, this final writing of temporal discontinuity employed the same imaginary conception of time as a punctuated, disjunctive process that we found in those accounts of new insecurity and new war. As with those narratives, this writing again emerged at the exclusion of alternative interpretations of its objects: refusing – most visibly – to acknowledge those left behind in this newly responsible and united American population. And, by hailing America(ns) into one united, collective self-identity, this construction of self-transformation also offered a potentially very productive discursive mechanism for creating continuing support for – and engagement in – this new War on Terror. For each of these reasons – both exclusionary and constitutive – this writing must again be viewed not as a neutral or inevitable reflection of those attacks and their aftermath. But, rather, and crucially, as a necessarily, inherently, political phenomenon.

Conclusion

As we have seen, claims to temporal break and discontinuity were integral to the writing of the War on Terror. By positioning 9/11 as a unique, exceptional, moment of historical rupture, the Bush administration was able to project the attacks' singular significance forwards into the future: locating this event as a dividing point between two discrete historical periods. Where the pre-9/11 era had been characterised by security, peace, and an individualistic American population, the post-9/11 period was one of insecurity, war, and social cohesion. What had hitherto seemed normal and expectable had been dramatically, and irrevocably, punctured by the four aircraft crashes of that morning. At home and abroad alike nothing would ever be the same again.

To return to the arguments introduced in Chapter 1 very briefly, reflecting on these three constructions of temporal discontinuity already takes us some way towards a further understanding of the relationships between time, identity, violence, and the political that we there pointed to. In the first instance, we have seen a number of ways in which self and other within this new war – or, perhaps better, selves and others within this new war – were brought into being through appeals to temporality: whether in the distancing of al-Qaeda and their equivalents from other earlier dangers, or in constructions of a brave new America, to offer but two examples. In the second, we have seen that the widespread impression of temporal break created by Bush's administration was also central not only to the legitimisation of particular violences – as war had already begun on 11 September 2001, military responses to this new threat should be expected. But, also, to the delegitimisation of particular violences, with claims to normative and discursive rupture, most visibly, working to strip off the 9/11 attacks from any sense that they may have possessed either political context or conceivable legitimacy.

Finally, we have also now begun our exploration of the political character of temporality itself in each of the senses that we sketched in our introductory chapter. In the first instance, as we have seen, claims to temporal discontinuity performed a vital constitutive role within the War on Terror's existence, status, and shape not least in

- problematising the contemporary terrorist threat as unique and unprecedented, thereby lacking in any historical context or political content;
- legitimising the institution of potentially controversial or exceptional responses to this new threat by calling for a congruity of exceptionality between problem and response;
- opening discursive space for expanding the War on Terror to a potentially infinite number of equally ambiguous unconventional security threats;
- justifying the institution of the War on Terror by denying the Bush administration any agency in this war's declaration;
- and introducing structural coherence into the War on Terror by depicting it as a new form of multifaceted conflict spanning many fronts simultaneously.

In the second, these writings also worked to conceal or exclude alterative writings of time, identity, and violence that could quite legitimately have been invoked within this particular context. Although interesting in its own right, this act of exclusion becomes even more urgent once we reflect on its productivity for foreclosing a whole range of alternative responses to the identities and violences that were both present and absent in this brave new war. And, as we shall now see in Chapter 4, assertions of rupture and punctuation were far from alone in performing either of these political functions.

4
Writing Linear Times

In the previous chapter we got our first insight into the significance and functions that claims to a privileged knowledge of time performed within the Bush administration's unfolding War on Terror. As we saw in that discussion, representations of 9/11 as a bringer of new insecurity, new war, and a new American public were integral to the framing of this conflict as an inevitable, necessary, and legitimate response to those (ostensibly unprecedented, exceptional) events. They were central, also, in opening space for subsequent expansions of this conflict beyond its initial targets of terrorism and those aiding or supporting terrorist organisations, and in introducing coherence into a multifaceted and dynamic discursive formation through facilitating the linkage of otherwise dissociable policies, developments, and political choices.

Given the repetitiveness with which Bush's administration turned to notions of temporal discontinuity in their describing and justifying this new War on Terror, it may seem surprising to posit the emergence of an entirely distinct – yet equally significant – conception of time within this conflict's unfolding. Yet, as we shall see in the pages that follow, that is precisely what happened in the earliest stages of this war. Predicated not on claims to historical rupture, breakage, or qualitative disjuncture but, rather, on assertions of continuity, evolutionary development, and progress, this second temporal shape of our discussion depicted the movement or flow of time as a far smoother, directional, linear dynamic. This chapter sets out to explore these efforts at writing the War on Terror around notions of linear time by again tracing three particularly prominent examples of its employment.

In the first instance, the following discussion introduces the persistent governmental attempt to write 9/11 not as a unique act of violence lacking parallel or equivalent. But, instead, as the latest, most deadly, assault within a lengthy chronology of terrorist and non-terrorist attacks against the United States and, indeed, others. Particularly interesting as a direct challenge to the dehistoricised, dislocatory framings of that event considered in the previous chapter, this writing positioned 9/11 as an entirely comprehensible (if not justifiable or directly predictable) evolution in unconventional warfare. And, as we shall see, although evident from the War on Terror's inception on 11 September 2001, this linear framing became increasingly prominent as this conflict continued.

The second construction of linear temporality explored in this chapter concerns the Bush administration's writing of the post-9/11 period as a time characterised by a gradual, if difficult and moderated, return to normality. With the American public here depicted as undergoing a slow, painful, return to the core values and beliefs they had long held, this writing offered a curious challenge to the sense of epiphany associated with the discontinuous framings considered in our earlier discussion of self-transformation. Of particular interest in this discussion, however, are the ways in which this radically different conception of time and identity offered precisely the same legitimatory functions for the War on Terror and its more controversial developments as did that with which we concluded Chapter 3.

In concluding this exploration of temporal linearity and the War on Terror, we turn finally to the Bush administration's seemingly unrelenting effort to present this conflict around a story of progress and incremental success. Predicated on a vision of the future as a knowable, foreseeable horizon, this writing again departed considerably from the claims to unknowable, unpredictable, times yet-to-come that we have been exploring in our discussion thus far. And, as noted in the previous chapter, I want to suggest here that this writing offered a particularly useful mechanism for (re)inscribing certainty and agency into this discourse. It did so, as we shall see, by offering a powerful sense of assurance regarding the War on Terror's necessity and ultimate triumph that was crucial for countering potentially unpopular setbacks and obstacles that would be – were being – encountered along the way.

By tracing the emergence and implications of each of these writings of temporal linearity, this chapter offers a further reflection on the relationship between time and the political with which we have been thus far concerned. For, as we shall see, claims to temporality as an ordered, stable, and progressive dynamic can be equally productive for political elites as claims to exceptionality, newness, and historical rupture.

Writing 9/11 as precedented continuity

In our previous discussion, we encountered the Bush administration's relentless effort to represent 9/11 as an exceptional, path-breaking moment in American, indeed global, political life. As already noted, that particular framing subsequently, and appropriately, received a great deal of academic interest and criticism. Although valuable in disturbing the sense of certainty with which those events were repeatedly positioned as a rupture of the normal, this interest has perhaps worked to camouflage a much distinct framing of those attacks that was also, if less frequently, invoked by the administration in their production of this discourse. To get a sense of the parameters and impact of this writing, let us return here to Francis Taylor's testimony to the Senate Committee on International Relations that we introduced in the previous chapter. For, having presented 9/11 as a rupture of the normal – an incomprehensible occurrence whose justification simply could not be discerned – Taylor's (25/9/2001) testimony moved quickly, and interestingly, to qualify this reading quite considerably:

> To summarize, in some ways the September 11 attacks do not reflect a brand new trend as much as a quantitative increase in the terrorists' sophistication, planning and willingness to cause large scale destruction and loss of life. ... The September 11 attacks were a continuation of the trend to inflict maximum casualties, without regard to loss of life or likelihood of achieving specific demands.

Taylor's positioning of 9/11 offers an explicit, and rather disconcerting, challenge to the writings of temporal discontinuity we encountered in the previous chapter. Where those events were depicted there as an indescribable trauma ushering in a qualitatively new time of insecurity, war, and selfish individualism, this framing relocates 9/11 into a palpably linear conception of historical change. Terrorism, we are now being informed, predated this particular, most recent, attack on America(ns). And, what had previously appeared to be markers of those events' very distinctiveness – strategic sophistication, large scale destruction, the production of mass casualties, and so forth – now appear to have characterised political violence for some time. If 9/11 did represent some form of transformation within terrorist tactics and determination, then, the change here appears neither radical nor abrupt. Instead, what we have is a story of change that is both quantitative and incremental and, as such, minor in relation to those accounts we have been discussing thus far. Situated, no longer, as a dehistoricised, decontextualised moment of rupture, 9/11 now emerges as but one – albeit significant – moment in a broader historical trajectory. And, as Taylor (25/9/2001) made clear in sketching the contours of this continuity, that particular historical trajectory stretched back for almost a decade:

> In the early 1990's, we saw the emergence of radical fundamentalist terrorist groups that relied not on state sponsors but primarily on funds raised independently through front companies and so-called charitable contributions. Unlike their predecessors of the 70's and 80's, these groups were distinguished by the fact that they were loosely knit international networks. Some had ties stemming from their involvement in the successful effort by the Afghan people to throw out the occupying forces of the former Soviet Union. It was from this group that Islamic extremist 'Afghan Alumni' formed the group al-Qaida, which means 'The Base' in Arabic. Al-Qaida is essentially a holding company comprised of many terrorist groups and independent cells. The President and CEO of this holding company is Usama Bin Laden, the 17th son of a wealthy Saudi businessman and veteran of the war in Afghanistan Bin Ladin's goal is to remove the American presence from Saudi Arabia and other Islamic nations and to create an Islamic utopia in what is now the Islamic world. He sees the United States as the major impediment to his goal and has vowed to attack America and Americans to undermine our influence on the world stage. While some attacks associated with al-Qaida were aimed against specific U.S. military targets, such as USS Cole in Yemen, others were aimed at civilians, such as the bombing of the World Trade Center in 1993 and the 1998 bombings of the U.S. embassies in Tanzania and Kenya that killed over 200 Africans, as well as 12 Americans.

In several important respects, this presentation of 9/11 parallels ongoing academic discussions concerning the emergence of a 'new' or 'postmodern' terrorism in the immediate post-Cold War era (Laqueur, 1996, 1999; Simon & Benjamin, 2002). These arguments identify a decline in state-sponsored, secular terrorist groupings at that particular historical juncture, and an attendant growth in religiously inspired organisations seeking ever more spectacular, ever more deadly, attacks against 'soft' Western targets. Whatever the accuracy of this (much-debated) thesis (see, for example, Tucker, 2001; Duyvesteyn, 2004; Burnett & Whyte, 2005), it is of particular interest for our current discussion in the problems it throws up for thinking 9/11 as some form of rupture. For, as Taylor's remarks vividly demonstrate, this linear writing of terrorist violence here posits those attacks as little more than the latest success of an increasingly destructive, increasingly incorrigible, threat to America(ns). A change from earlier attacks, in this writing, to be sure. But, here, a foreseeable, comprehensible change: at least with the benefit of political hindsight.

Despite the catalogue of precedents noted by Taylor, this alternative positioning of 9/11 at the cusp of a pattern of unconventional violence was rather less visible than the discontinuous writings of those events: at least in their most immediate aftermath. As the War on Terror gained momentum, however, something appeared to shift in discussions of their historical location, and the Bush administration began to invoke this linear writing with ever-increasing regularity. On numerous occasions, as in Taylor's earlier statement, this invocation involved the identification of a whole series of earlier – albeit primarily recent – terrorist attacks on the United States. Not as violences so radically distinct from the uniqueness of 9/11, but rather as genuine precursors to this (now) latest occurrence (see, for example, Bush, 29/10/2001; Bush & Bondevik, 5/12/2001; Prosper, 20/2/2002; Rice & Ridge, 9/10/2001; Rice, 16/5/2002; Rumsfeld, 9/9/2004). In Taylor's (23/10/2001) subsequent remarks also before the Senate Committee on International Relations, for example, he returned once again to this writing of 9/11 as symptomatic of a continuing affliction confronting America(ns):

> Terrorist bombings and similar attacks have been a particular scourge in recent years. The United States has unfortunately been a frequent victim. In addition to the tragic events of September 11, the 1993 World Trade Center attack, the 1996 attacks on the U.S. Khobar Towers facility in Saudi Arabia, the 1998 attacks on our embassies in Kenya and Tanzania, and the October 2000 attack on the USS Cole, triggered both international condemnation and multinational cooperation in bringing perpetrators to justice.

As in his earlier comments, Taylor here again strips 9/11 of both its originary force and dislocatory mystique. Presented as but the latest example within a traceable chronology of victimhood and response, any sense of exceptionality accompanying this most recent attack is lost in this framing. The terrorist threat facing America, it seems, did not begin on the morning of 11 September 2001. Nor, interestingly, did the international condemnation of unconventional violence or the global determination to confront this (now less new) security challenge.

As the War on Terror's longevity stretched into months and years, this linear writing of the 9/11 attacks appeared to gain currency. With those events' second anniversary approaching, for example, John Ashcroft (9/9/2003) sought to follow Taylor's example in drawing a linkage between 9/11 and the earlier 1998 bombings of American embassies in East Africa:

> Al Qaeda had previously demonstrated the ability to strike again; to carry out multiple, complex and simultaneous attacks. The 1998 bombings of the U.S. embassies in Kenya and Tanzania were virtually simultaneous. 224 people lost their lives in explosions that occurred four minutes and 450 miles apart. And al Qaeda already had demonstrated the will to strike again. The same year bin Laden bombed our embassies in Africa, he ordered his followers in al Qaeda to murder innocent Americans wherever and whenever possible.

With the simultaneity, complexity, and destructive character of 9/11 appearing, once more, rather than less unique, Ashcroft again positions those events within a patterned, discernible trajectory of earlier strikes by al-Qaeda. Bin Laden's unrelenting determination to murder innocent Americans, it seemed, neither arrived on 11 September 2001 nor ended that day. And, seen through this lens, the post-9/11 security climate quickly took on a quite different hue to the new world of ever-present, ever-imminent, danger that we encountered in the previous chapter:

> So pressing has been the danger of further terrorism, and so enormous the task of protecting Americans from terrorism, that from time to time I doubted whether we would commemorate the second anniversary of September 11 without seeing additional loss of life and liberty on American soil. At times, I doubted America could make it, still safe, still secure, still free, to this day. That we have had two years of safety is first a blessing, a sign that God's grace continues to shine on our nation and its people.
>
> (Ashcroft, 9/9/2003)

To get a clearer sense of the growing significance of this linear writing to which I alluded above, it will be valuable briefly to point to its continuity even further beyond the events of 9/11. As the end of 2005 drew near, for example, with the violence in Iraq showing unfortunately limited signs of genuine decline, the Vice President drew on this framing at a Rally for the Troops held at Al-Asad Air Base, arguing:

> The terrorists' war against America began long before 9/11. And during those years, they were the ones on the offensive. They grew bolder in their belief that if they killed Americans, they could change American policy. In Beirut in 1983, terrorists killed 241 of our servicemen. Thereafter, the U.S. withdrew from Beiruit. In Mogadishu in 1993, terrorists killed 19 Americans; thereafter, the U.S. withdrew from Somalia. Over time the terrorists concluded that they could strike America without paying a price because they did – repeatedly: the bombing of the World Trade Center in New York in 1993; the murders at the Saudi National Guard training facility in Riyadh in 1995; the attack on Khobar Towers in 1996; the attack on our embassies in Kenya and Tanzania in 1998; the attack on the USS Cole in 2000. Ultimately, of course, they attacked the homeland on 9/11 and took the lives of 3,000 people aboard passenger jets, and at the World Trade Center, and at the Pentagon.
>
> (Cheney, 18/12/2005)

While, more recently still, President Bush (19/4/2007) traced a similarly explicit linkage between 9/11 and earlier attacks on American citizens in a 2007 discussion at Tipp City, Ohio:

> I realized [on 11 September 2001] that there is an enemy of the United States that is active and is lethal. At further study of that enemy, I realized that they share an ideology, that these weren't – that the – and when you really think about it, the September the 11th attack was not the first attack. There was a 1993 World Trade Center attack, there was attacks on our embassies in East Africa, there was an attack on the USS Cole, there have been other attacks on U.S. citizens, and that these attacks were instigated and carried out by cold-blooded killers who have a belief system. They are threatened by free societies. They can't stand the thought of freedom being the prevailing attitude in the world because their view is, if you don't believe in what I believe in, you probably shouldn't be around.

Revisiting, here, his earlier State of the Union claim that 'In the mind of the terrorist, this war began well before September the 11th, and will not end until their radical vision is fulfilled' (Bush, 23/1/2007),

9/11 emerges once again as a product of the very same, if increasingly emboldened, terrorist threat that had been confronting America(ns) for a number of years. As the *2003 National Strategy for Combating Terrorism* had previously noted: 'Americans know that terrorism did not begin on September 11, 2001. Regrettably, its history is long and all too familiar' (White House, 2/2003: 7–8).

We shall have more to say of the significance of this construction of 9/11 around a conception of temporal continuity in the discussion that follows. Before that, however, we must first turn to an important expansion of this particular writing. For, not only were these recent attacks inserted into a (knowable, directional) history of assaults on America by the Bush administration in their production of the War on Terror. They were also, at times, inserted into a distinctively less parochial history. Consider, for example, Powell's (21/9/2001a) statement to the Special Session of the Organization of American States:

> This is not the first time that nations of our hemisphere have suffered at terrorists' hands. The United States has stood with you and now you stand with us, partners in resolve as well as in grief. Free peoples committed to the collective defense of our security and of the democratic ideals that we hold so dear.

On first perusal, Powell's remarks simply replicate those we have considered thus far. For, once again, he appears here to question 9/11's exceptionality by inserting those attacks again into a discernible – if regrettable – history of unconventional violence. Interestingly, however, this history of terrorism is here not an American history alone: the experience of a nation targeted for assault because of its particular (democratic) values and ideals. Rather, what we have now is instead a collective history forged around the unfortunate yet shared experience of a far broader – hemispheric – subject of violence. In this sense, where, in the previous chapter, we saw those attacks viewed as ushering in a new collective present/future of insecurity and threat. We have here the location of those events into a collective past far transcending the American borders. And, as Bush's subsequent (11/3/2002a) remarks at the White House South Lawn suggested, this collective subject of violence could be expanded still further when the occasion permitted: 'September the 11th was not the beginning of global terror, but it was the beginning of the world's concerted response.' Although Bush, interestingly, combines 9/11's linearity here with a discontinuous account of the response to that attack, this most recent event is once more, at least partially, stripped of its originary, rupturing power.

Of all of those within the administration capable of discussing recent historical violences against America(ns) and others, Donald Rumsfeld – as head of Defense – was perhaps most conveniently located in this discourse. As such, it is fitting, then, to end our discussion of this writing by turning to his own willingness to invoke this global history of terrorism: a global history of terrorism both pre-dating and following the 9/11 attacks. As he suggested in a 2004 interview with WHO-AM, for example:

> Well, if you think about it, the Russians lost hundreds and hundreds of children in terrorist attacks last week. We lost 3,000 three years ago this week. We've had terrorist attacks in Turkey, Tunisia, Indonesia, Morocco, country after country over many years. We lost American lives in the U.S.S. Cole. We lost American lives in the embassy bombings in East Africa. Osama bin Laden declared war on the United States back in the late 1990s.
>
> (Rumsfeld, 7/9/2004)

A theme he returned to just four days later in a wreath-laying ceremony at Arlington Cemetery to commemorate the third anniversary of the events of 11 September 2001:

> Those we mourn today were not the first victims in the war declared against us by the extremists, nor were they the last. All across the world, the enemy has taken innocent lives. Extremists think nothing of cutting off peoples' heads to advance their cause. They have murdered citizens – even hundreds of schoolchildren recently – from countries across the globe. And even today they plot to strike again.
>
> (Rumsfeld, 11/9/2004)

Even more recently still, finally, Rumsfeld offered a veritable catalogue of extremist attacks against both Americans and people of other nationalities in his March 2006 remarks at Carlisle Barracks, Pennsylvania:

> Though we think of September 11th as the first day in the Global War on Terror, it wasn't the first day for the enemy. Extremists had declared war on free people decades ago. In 20 years terrorists attacked and killed Americans more than 20 times including the bombing at: The U.S. embassy in Beirut, Lebanon, in 1983; The Marine barracks in

> Beirut in 1983; Pan Am Flight 103 over Lockerby, Scotland, in 1988; The New York World Trade Center the first time in 1993; A military compound in Riyadh, Saudi Arabia, in 1995; Khobar Towers in Saudi Arabia in 1996; U.S. Embassies in Kenya, Tanzania in 1998; and then the warship USS Cole in Yemen in 2000. During those decades the West was ambivalent about how to counter extremist ideology and that type of aggression. As a result, terrorists became increasingly bolder. We should have learned the timeless truth – that weakness is provocative. ... On September 11th, they accomplished their most daring attack on our shores, and in the years since no part of the world has really been spared from their attacks.
>
> (Rumsfeld, 27/3/2006)

As with the earlier accounts, we have here a quite discernibly and radically distinct production of 11 September 2001 to those we have been exploring thus far.

Political precedents

As we have seen, official discussions of the 9/11 attacks that took place within the rubric of the unfolding War on Terror were far from limited to the discontinuous accounts we encountered in the previous chapter. As the above examples so lucidly demonstrate, Bush, Powell, Rumsfeld, and many of the other most high-profile administration representatives were keen to position 9/11 not (only) as an exceptional, uniquely destabilising event divorced from any meaningful historical context. But (also), as the successor to any number of earlier assaults: both in America and, indeed, far beyond. Although quantitatively distinguished from its precedents with invocations of this attack's greater destructiveness, 9/11 now emerges as part of a discernible, linear, continuity of terrorist violences: a discernible, linear, continuity of terrorist violences, indeed, both pre-dating and subsequently post-dating those events. And, with temporality reimagined here as a directional, continuous, and traceable dynamic, this linear writing quite clearly offers a considerable challenge to the sense of dislocatory, tragic, mystery we saw being inscribed into 9/11 in our earlier discussion. For, in the very attempt to connect 9/11 with earlier and later occurrences, those events now re-emerge as possessing those very qualities we encountered their having there appeared to have lacked. 9/11 may, in this construction, indeed be described, understood, and historically located – however imperfectly: no sense of uniqueness, no sense of newness, prevents our contextualising this violence here. And, let us not forget, finally, that this alternative writing also offers a quite considerable challenge to the powerful sense of instituting agency we earlier saw assigned to those events as the bringer of War on Terror. For, as many of the above statements explicitly note, this particular war could be traced far back beyond the events of that morning should we so choose to read temporality through a rather different lens.

In one sense, it would be tempting to approach this linear writing as a more accurate – or, perhaps better, fuller – account of 9/11 than those considered in our earlier discussion. Unconventional, non-state, violences against the United States and others had, clearly, occurred prior to this most recent occurrence. However limited the continuity between all of the above-referenced events, it is certainly possible to make the case that they seemed to share something and, with that case having been successfully presented, to begin to connect them around a logical structure of beginnings, middles, and ends. For reasons I would like now to suggest, however, I think we need to resist the appeal to proclaim this writing a more accurate or better account of the 9/11 attacks than the discontinuous presentations we have studied thus far. For, as with those accounts of rupture and qualitative change, this alternative positioning must also be viewed as a contingent, exclusionary, and hence political production on several grounds.

Perhaps the most obvious and important exclusion of this linear writing centred on its attempt to insert 9/11 into a *single*, unidirectional, history of terrorist violence. By connecting this particular attack to others as temporally, spatially, and contextually distinct as the 1983 bombing of the U.S. embassy in Beirut, the 1988 Lockerbie bombing, the 1993 World Trade Center bombing, and the more recent 2004 incidents in Beslan, this writing offered very limited space for acknowledging the distinctions between any of these events and the others noted above. Whatever we may think of these occurrences, it seems remarkably unproductive to subsume them under the rubric of a singular trend: a singular history of equivalent – if ever-more dangerous and sophisticated – terrorist incidents. With the agents, motives, and logistics of these attacks so incredibly different, this appeal to one History – one 'Terrorism with a capital T' (Aretxaga, 2002: 145) – appears particularly dubious. That political violences had, no doubt, occurred in the past, should not here justify our running roughshod over the differences separating these violences. For, as Post (2005: 616) has recently noted:

> [I]t is important to consider each terrorism in its own political, historical, and cultural context, for terrorism is a product of its own

> place and time. It is an attractive strategy to a diverse array of groups which have little else in common. ... So we should be discussing terrorisms, plural – not terrorism – and terrorist psychologies, plural, rather than searching for a unified general theory explaining all terrorist behavior.

Post's remarks inject a welcome note of caution into the above readings of 9/11's historical location and context. It is, of course, eminently preferable to approach political violences in the specificities of their occurrence: to recognise and explore heterogeneity amidst claims to equivalence, whatever the object of analysis under discussion. Importantly, however, this caution only takes us so far, in that one further exclusion of this linear writing remains as yet concealed. For, not only did this positioning of 9/11 militate against any sensible consideration of context or background that may have been pertinent to understanding events in Beirut, Lockerbie, Beslan, and beyond, it also militated against any sustained reflection on the possible legitimacies or justifications of those violences themselves. By presenting 9/11 as but the most recent cold-blooded murder of innocents – but the most recent in a long, tragic, history of amoral assaults – this linear writing rendered any discussion of the morality of those assaults redundant and void. Not only were these events *equivalently* terrorist in this positioning, they were also, crucially, equivalently *terrorist*. In this sense, although offering a rather different conception of time's movement and flow to those we encountered in the previous chapter, this production effectively complemented those discontinuous writings by further depoliticising the recent attacks on America(ns), further demonising the agents behind them. Although many will, no doubt, have little problem with this condemnatory rhetoric – approaching political violence as necessarily, absolutely, morally impermissible whatever its context or motives – surely it would be preferable to offer a detailed analysis of the causes, alternatives, and consequences of each of these violences away from the appeal to a universal morality underpinning this blanket act of demonisation.

Before turning to the second writing of temporal linearity that concerns us in this chapter, I want to conclude our discussion here by turning to the productivity of this particular construction of continuity and historical lineage. For, where we saw in the previous chapter that representations of discontinuity worked to inscribe significance into 9/11 by appealing to its unprecedented exceptionality, this linear positioning achieved something remarkably similar – albeit in a very different way. By positioning 9/11 at the cusp of a continuing and escalating pattern of violence, this construction rendered the Bush administration capable of pointing to a long history of American inaction and/or ineffectiveness as a justification for this war's political and strategic necessity. A long history that became particularly useful, indeed, as controversies over this war's continuation grew ever more loud. As Cheney (18/12/2005) argued in his 2005 'Rally for the Troops in Iraq' noted above: 'Over time the terrorists concluded that they could strike America without paying a price because they did – repeatedly.' Having (now) witnessed the futility of earlier counterterrorist strategies centred on either the limited use of military force:

> If you think back with me for a moment to 1983, 20 years ago, the attack on our Marines in the barracks in Beirut, Lebanon, 241 killed one morning, truck bomb, probably Hezbollah. There was no effective U.S. response to that attack. A few months later, we withdrew, pulled out of Lebanon, brought everybody home. 1993, World Trade Center bombing; 1995, attack on Saudi Arabian National Guard Headquarters, five American advisors killed; '96, Khobar Towers bombing in Saudi Arabia; '98, East Africa embassies, two of our embassies hit simultaneously, hundreds killed, 12 Americans; in 2000, the USS Cole struck off Yemen, 17 sailors killed – you ask yourself what was the U.S. response to those attacks? How did we deal with the fact that we'd been struck? And the answer is, we did almost nothing – launched a few cruise missiles once or twice at some empty camps in Afghanistan. But that was really the sum total of the U.S. response.
>
> (Cheney, 3/10/2003)

Or, indeed, on traditional law enforcement measures:

> The President properly determined after September 11th that the United States no longer could deal with terrorists killing our people as we had in the past in the traditional law enforcement sense. Indeed, the only way to defeat terrorism is to go on the attack. That's exactly what our coalition has done and is doing in Afghanistan, Iraq and elsewhere around the world.
>
> (Rumsfeld, 2/8/2005)

Aggressive, determined, and continuing military action emerged in this production as the only viable route to genuine security. For, as President

Bush (20/4/2007) argued more recently in the face of continuing calls for the withdrawal of American troops from Iraq:

> The lesson of 9/11 is that when you allow extremists and radicals and killers to find a sanctuary anywhere in the world, that can have deadly consequences on the streets of our own cities. ... Withdrawal would embolden enemies and confirm their belief that America is weak and does not have the stomach to do what is necessary to lay the foundations for peace. Ultimately, withdrawal would increase the probability that American troops would have to return to Iraq – and confront an enemy that is even more dangerous.

In short, by positioning 9/11 at the cusp of a (knowable and traceable) linear trend of terrorist activity, this writing worked to further (re)inscribe strategic significance and necessity into the War on Terror itself. Not by appeal to a radically new security environment into which America(ns) had been recently, and dramatically, catapulted – for terrorism, in this writing, had been around for a very long time, but, rather, by virtue of the historical lessons learned in earlier confrontations with essentially the same enemy that was confronting America(ns) and their allies today.

Beyond its more directly legitimatory functions, this first linear writing also offered an important contribution to the Bush administration's efforts at writing the War on Terror as a truly global conflict: a conflict that could not but be of interest to nations and peoples far beyond the American borders. By positioning 9/11 within an ostensibly coherent chain of events stretching not only backwards into the past but also forwards into Turkey, Tunisia, Indonesia, Morocco, and even Russia, the administration was able to locate the War on Terror as an emphatically transcontinental enterprise and concern. This construction of a global collective identity with a stake in this conflict not only added further moral legitimacy to the unfolding war by tapping (perhaps imperfectly) into the just war tradition's concern with the need for legitimate authority (Jackson, 2005: 130–2; Rengger, 2002: 358). It also offered the administration a potentially very productive discursive device for obtaining the continuing support of other states for their war. For, failure to arrest this (constructed) evolutionary path would threaten not only American but also global security in this writing. As we (the hemisphere, civilisation, the world) had experienced terrorism before, so we (the hemisphere, civilisation, the world) must take action not to experience it again. In this sense, and paralleling those constructions of new insecurity and new war considered above, this first linear narrative functioned as a global interpellative call to act against a common (ever-expanding) concern.

To summarise briefly, the above section has explored the Bush administration's persistent, if initially less visible, attempt to write 9/11 as a quantitative, linear, evolutionary development from earlier terrorist attacks. This writing, as we have seen, was particularly interesting in its introduction of an explicit ambiguity into the above constructions of unprecedented and indescribable rupture: locating the attacks now within a discernible, knowable historical trajectory. While predicated on the exclusion of alternative writings of that history of violences, it offered a particularly productive discursive resource for inscribing significance into the unfolding War on Terror, while also opening space for expanding this war beyond the United States' borders. And, as such, it should be viewed again as an inherently political construction in the senses of the term outlined in the account offered in Chapter 1.

Writing return to (moderated) normality

The second official representation of temporal linearity that emerged in the aftermath of the 9/11 attacks related to the Bush administration's frequent writing of that period as a time of return to moderated normality. In direct contrast to the writing of self-transformation with which we concluded the previous chapter, this particular construction located those attacks not as a 'watershed' moment that had so radically, so rapidly, transformed the lifestyles, values, and beliefs of the American public. Rather, it depicted the aftermath of those events as characterised by a gradual, if difficult and qualified, *return* to earlier values and beliefs: a return (necessarily, and understandably) modified by a new understanding of those values' precariousness. By locating an incremental, evolutionary change in the American population, then, this writing mobilised a similarly linear conception of temporality to that considered above: a conception of temporality characterised by moderated continuity and emphatically not rupture.

Consider, to begin with, Colin Powell's 12 September 2001 on-the-record briefing for the press:

> Now, yes, we believe that acts of war have been committed against the American people, and we will respond accordingly. But at the same time, life has to go on. In all of the difficult times we will be facing ahead, we have to still try to return life to a sense of normalcy.

> We cannot be a people who are afraid to live. We cannot be a people who will move away from a relatively open society. We cannot be a people who walk around terrified. We're Americans; we don't walk around terrified. We are going to be strong in this difficult period, and we're going to move forward with pride and with determination. And we will get our society back to normal with whatever additional precautions, nevertheless, might be necessary to secure our society without locking ourselves down.
>
> (Powell, 12/9/2001a)

This statement offers a markedly different interpretation of the American population's likely response to the 9/11 attacks than that underpinning the writings of self-transformation considered in the previous chapter. Although speaking but one day after this (otherwise ostensibly unspeakable, traumatic) event, Powell is already here confident of two future occurrences. First, he is strikingly certain that the post-9/11 period will witness a time of recovery within the United States – 'we will get our society back to normal': a return to normality *in spite of* the recent acts of war. And, second, he is also explicit in suggesting that this return to normality may be limited or moderated – 'with whatever additional precautions ... might be necessary': an unfortunate, if necessary, period of adaptation to ensure the continuation of the 'relatively open society' long enjoyed by America(ns). For, as he had earlier argued that day: 'this is a time for caution and vigilance. But it is also a time for us to get back to work' (Powell, 12/9/2001b). If America was to be changed by the attacks of 11 September 2001, then, the character of that change would here be rather more minimal and quantitative than the moment of epiphany encountered in Chapter 3.

Powell returned to this second writing of temporal linearity in his subsequent, 21 September 2001, remarks with the Canadian Minister of Foreign Affairs John Manley. As with each of his earlier statements, these remarks once more attributed an essential continuity to American society and its inhabitants: albeit a continuity moderated again by necessary changes:

> Well, I think our lives have changed. ... But it's also important to remember that we are a people who live in an open society and we don't want the society to become closed. We need people to go back out to stores. We need people to go to movies and theaters. We need to restore a sense of normalcy in our life, while at the same time being mindful of the challenges to our security that exist. But the

> terrorists will really have won if they have changed our fundamental way of life. And they won't do that, they can't do that. But at the same time, we have to show an added level of security in order to protect ourselves and to protect our citizens.
>
> (Powell, 21/9/2001b)

If extra security measures still remained necessary to protect and preserve American society, the core, fundamental, openness of that society neither would be, nor should be, changed here by the terrorists' actions. And, as he had earlier argued, indeed, this return to moderated normality would also be participated in by Americans living far from their country of origin:

> Our embassies are all hard at work. There have been some closures, and they go down and come back up in response to threat conditions, but our plan is to be actively engaged around the world and not let this heightened sense of tension affect our ability to do our job. And so we are encouraging all of our ambassadors to do smart things, to take all necessary safety precautions and to make sure their security is intact, but at the same time to continue doing America's business throughout the world.
>
> (Powell, 13/9/2001)

As with each of the writings considered thus far, this second, alternative, reflection on the American public's experiencing a qualified return to the normal found resonance throughout many of the core sites of the Bush administration. From the Department of Justice, for example, Deputy Attorney General Larry Thompson proved a particularly keen advocate of this linear construction: invoking this directional model of time frequently in explicit juxtaposition to the writings of radical discontinuity considered above. In his November remarks at Western Fairfax, Washington, for example, Thompson (8/11/2001) suggested:

> Since September 11, I have heard from a number of people about the difficulty we all have experienced getting back to what we once called 'normal'. We can find no acceptable 'routine', it seems, while memories of the terrible events of September 11 remain clear. In spite of this, however, I have noticed something remarkably normal in our common reaction to this great tragedy; I'm sure you've noticed, too. It's spirit – that American spirit that's so difficult to define but so easy to recognize when things get tough. We saw this spirit in the

> response of the police and fire departments and emergency medical teams at the crash sites in New York, Virginia, and Pennsylvania. We see it in the outpouring of help in money and material donations, in the long lines of blood donors, in the many, many community volunteer initiatives to coordinate relief for victims of the attacks. I have experienced this personally in the many, many phone calls and e-mails I've received from those of you in private practice asking what you could do in helping to address the events of September 11. We recognize this as a normal American response to any abnormal event – we get together and help the victims, we reach out to fellow Americans and hold our liberties closer.

As in Powell's earlier remarks, despite the (understandable) transformations in our (American) routines engendered by this abnormal event, a continued (and emphatically not new) spirit of sacrifice, liberty, and compassion clearly here characterises post-9/11 America. His (near-verbatim) return to this writing in a 15 March 2002 speech at the American College of Trial Lawyers (Thompson, 15/3/2002) only prefaced his subsequent, 2 May 2002, dismissal of claims that contemporary security measures would 'change the essential character of our nation':

> Americans before us have gone to war many times – and spent untold blood and treasure to secure the liberties that we hold dear – but we are the first generation of Americans to confront the mass murder of civilians by a foreign enemy on our own soil. As a public official – and as an attorney who, for many years, represented criminal defendants – I share the concern that the struggle against terrorism not change the essential character of our nation. But I want to assure you that none of these steps threatens our Constitution, our essential liberties. Despite the gravity of the terrorist attacks and the threat of future attacks, all of the measures that we are bringing to bear have been arrived at openly, in the sunlight of public attention, and are subject to judicial review. It is our very open, democratic and just society – whose hallmark is our concern for civil rights – that has made us the terrorists' target. ... This same reverence for civil rights inspired the Attorney General immediately after the terrorist attacks to challenge all of us in the Justice Department to 'think outside the box' in fighting terrorism, but caution us: 'Don't think outside the Constitution.'
>
> (Thompson, 2/5/2002)

Whatever 'outside the box' changes were to occur in American society, written thus, the fundamental core of that society would continue to remain protected by the checks and balances of democratic norms and procedures.

Although many within Bush's administration employed this second writing of linear temporality within the initial stages of the unfolding War on Terror (see, for example, Ashcroft & Mueller, 20/9/2001; Cheney, 25/10/2001; Ridge et al., 30/10/2001; Ridge, 27/11/2001), its most prominent advocate was often the President himself (see, for example, Bush, 2/10/2001a, 3/10/2001, 4/10/2001a, 12/10/2001, 11/3/2002b). Consider, for example, his 15 September 2001 response to a question concerning the sacrifices Americans could be expected to make in their daily routines following 9/11:

> Our hope, of course, is that they make no sacrifice whatsoever. We would like to see life return to normal in America. But these people have declared war on us and we will do whatever it takes to make sure that we're safe internally. So, therefore, people may not be able to board flights as quickly. Our borders are tighter than they've ever been before. We've taken a variety of measures to make sure that the American people are safe, just as the Attorney General spoke about. But we hope, obviously, that the measures we take will allow the American economy to continue on. I urge people to go to their businesses on Monday. I understand major league baseball is going to start playing again. It is important for America to get on about its life. But our government will be on full alert and we'll be tracing every lead, every potential to make sure that the American people are safe.
>
> (Bush et al., 15/9/2001)

As in each of the earlier statements, this writing once again posits a post-9/11 evolutionary transformation in American society. If future security precautions ('tightening our borders'; 'tracing every lead') were understandable and necessary in the wake of this 'declaration of war', American (here financial and sporting) life would effectively continue unabated.

Bush (2/10/2001b) returned to this theme of altered continuity in his subsequent, 2 October 2001, meeting with Congressional Leaders in the Oval Office:

> I say that America ought to be on alert, but we need to get back to business. That's why I'm opening up Reagan Airport. That's why we

had Cabinet members get on commercial airlines over the weekend. The good news is, is that some of the load factors on American airlines looked like they increased over the weekend. Americans know their government is doing everything they can to disrupt any terrorist activity that may occur. We're following every lead, we're interrogating every possible suspect. We're on full alert in America. But the good news is, Americans also realize that in order to fight terrorism, they're going to go about their lives in a normal way. And Americans are.

And, again, in his, 9 October 2001, remarks with Chancellor Schroeder, wherein he positioned both 9/11 and the recent anthrax attacks as equally ineffective in bringing any fundamental transformation to American society:

> [A]ny type of incident, any type of information that comes into our government, we take very seriously. Because we understand we're dealing with evil people. And as you know, I've assured the American people that life should go on as normal. But they also should know, the American people should know, that our government is doing everything we can to make our country as safe as possible. ... But the American people should go about their business. And people ought to travel. And people ought to feel comfortable going to see the mighty Houston Astros play. People ought to feel comfortable going about their lives, knowing that their government is doing everything humanly possible to disrupt any potential activity that the evil ones may try to inflict upon us.
>
> (Bush & Schroeder, 9/10/2001)

His 11 October 2001 news conference similarly positioned the attacks thus, with the suggestion, 'We are getting back to normal. We're doing so with a new sense of awareness' (Bush, 11/10/2001b); as, finally, did his 2002 State of the Union Address, within which he argued:

> The last time I spoke here, I expressed the hope that life would return to normal. In some ways, it has. In others, it never will. Those of us who have lived through these challenging times have been changed by them. We've come to know truths that we will never question: evil is real, and it must be opposed. Beyond all differences of race or creed, we are one country, mourning together and facing danger together. Deep in the American character, there is honor, and it is

> stronger than cynicism. And many have discovered again that even in tragedy – especially in tragedy – God is near.
>
> (Bush, 29/1/2002)

In short, as had other representatives of his administration, Bush persistently located an essential core continuity inhabiting American society and its people. In spite of the new sense of awareness and accompanying changes that understandably characterised that society, something important would remain largely unaffected: something would continue in spite of the recent trauma experienced on 11 September 2001.

The politics of returning to normal

As with the writing of 9/11 as precedented continuity, the Bush administration's discussion of the post-9/11 period as a return to moderated normality employed a fundamentally distinct conception of temporality to those considered in the previous chapter. Substituting a time of discrete, heterogeneous, periodicity with one of linear directionality, this writing constructed American society as a site of relative persistence. Although adjustments in the domestic security architecture would no doubt take place in the aftermath of the 9/11 attacks, those adjustments were necessary to ensure the continuation of the core values and practices that had long characterised America. Not a radical transformation in domestic social and political life, then, but, rather, a development that would remain faithful to embedded and established behaviours and principles. Business, professional sport, and all of the other trappings of that relatively open society would continue – were continuing – in spite of the best efforts of those opposing the American way of life. Airports were reopening, Major League Baseball was being relaunched, and employees were returning to their workplaces. If the 9/11 attacks had brought about minor changes within the United States, these changes had emphatically not transformed it as the terrorists had hoped: instead they had only forced the consolidation of an existing lifestyle. In this sense, we have here a fundamentally different scripting of America(ns) than that offered in the final construction of our previous chapter. America(ns), here, had not shed their shallow, materialistic lifestyles in a moment of epiphany brought on by the collective trauma inspired by 9/11. They would shop as they had done before, labour as they had done before, leisure as they had done before, while accepting, at the same time, any and all necessary security precautions that would accompany them in these pursuits. Not as restrictions on the lives they were hoping to lead. But, instead, as essential enabling conditions without which any semblance of (moderated) normality could simply not be achieved.

What is particularly interesting in this second linear writing is that, although appearing to offer a strikingly different account of American society to the transformatory writing of our earlier discussion, these two constructions shared one important exclusionary moment. As outlined in Chapter 3, the construction of a new America recently united around self-sacrificial ideals and so forth offered remarkably little space for acknowledging those members of the public who may legitimately have felt unrepresented in these claims to a brave new world. For, as we there noted, the targeting of specific minority populations for their perceived responsibility or complicity in the attacks, for example, did not appear to sit particularly comfortably with this construction of a newly harmonious spirit of social cohesion.

In a similar sense, this linear writing of moderated social return also worked to conceal the experiences of those individuals not privy to the trappings and benefits of the open normality attributed to American society: either before the 9/11 attacks or, indeed, afterwards. Once we begin to take those experiences into consideration, the sense of continuing – albeit modified – liberty and openness accompanying the above discussions begins to look at best selective, at worst simply deceitful. Most obviously, as David Cole (2002: 957) has argued so forcefully, non-citizens in particular had good reason to feel unrepresented in this particular story, where:

> Since September 11, immigrants have been the targets of a massive preventive detention campaign conducted under unprecedented secrecy. Congress has enacted new laws that subject noncitizens to guilt by association, ideological exclusion, and unilateral executive detention. And federal and local officials have engaged in ethnic profiling, treating immigrants as suspect based on little more than their Arab ethnicity or national origin.

However unsavoury the impact of these new security measures on noncitizens, those individuals were not, of course, the only ones affected by this new security environment. Legitimised under a dubious, metaphorical, rhetoric of balancing liberty and security (compare Waldron, 2003 with Meisels, 2005), the Bush administration moved quickly to introduce a whole sweep of increasingly invasive instruments of control under a new doctrine of homeland security (see, for example, Ahmad, 2002; Gould, 2002; Rackow, 2002; Baker, 2003). Although typically presented in universalist rhetoric, there is little doubt that these instruments impacted on particular groups, particular communities, of Americans quite differently. As the Migration Policy Institute (2003: 9) concluded in its 2003 report *America's Challenge: Domestic Security, Civil Liberties, and National Unity after September 11*, for example, the furtive arrest and detainment of over 1200 individuals following 9/11 had 'terrified and alienated' Arab and Muslim communities across the United States. Seen through this lens, then, this writing of a singular, unitary American society returning to moderated normality appears once again as an exclusionary political construction: one that could only function by negating the experiences of those individuals (either newly or perpetually) located on the margins of American society.

In our previous chapter, we encountered a number of key ways in which governmental writings of 9/11 as historical discontinuity worked to legitimise potentially controversial responses to those events. Where those constructions of rupture offered political legitimacy to this war by appeal to a new and exceptional security environment that had been entered that day, this – very different – writing performed a similar function, albeit in a rather distinct manner. For, as we can see by reconsidering Thompson's 2 May 2002 reflections on the War on Terror's domestic front, it did so by *disavowing* the very exceptionality of the measures introduced under this rubric:

> As a public official – and as an attorney who, for many years, represented criminal defendants – I share the concern that the struggle against terrorism not change the essential character of our nation. But I want to assure you that none of these steps threatens our Constitution, our essential liberties. Despite the gravity of the terrorist attacks and the threat of future attacks, all of the measures that we are bringing to bear have been arrived at openly, in the sunlight of public attention, and are subject to judicial review.
>
> (Thompson, 2/5/2002)

As Thompson's remarks quite explicitly suggest, this linearised writing of the unfolding war allowed the Bush administration to defend these new security measures as necessary and effectively peripheral mechanisms for engendering the continuation of essential American liberties: for engendering the continuation of essential American liberties *in spite of* the continuing threat of future attacks. Presented in this way, these measures were no longer characterised as an exceptional response to exceptional circumstances. Rather, they were moderate, democratically sanctioned, guarantors of a critical – and therefore far more significant – core way of life. For, as Bush had earlier argued on signing the 2001 USA PATRIOT Act:

> This bill was carefully drafted and considered. Led by the members of Congress on this stage, and those seated in the audience, it was crafted with skill and care, determination and a spirit of bipartisanship for which the entire nation is grateful. This bill met with an overwhelming – overwhelming agreement in Congress, because it upholds and respects the civil liberties guaranteed by our Constitution.
>
> (Bush, 26/10/2001)

Finally, the constitutive, productive, impacts of this second representation of temporal continuity also, interestingly, paralleled one further implication noted in our previous discussion of the writing of American self-transformation. Where that account of a newly created American public worked as a productive disciplinary demand for social cohesion within the United States, this construction offered something similar. Not only did it work to exclude any meaningful recognition of those individuals or communities not partaking in this return to normality. It also coerced against dissenting readings of this unfolding war as an abnormal or exceptional development in established security practices at home. For, as we have seen, measures such as the USA PATRIOT Act were to be read now not as a unique development rendered necessary through confrontation with an equally unique, equally abnormal, threat to America(ns), but, rather, as congruent with – as continuation of – earlier efforts at combating or managing threats. In this sense, this writing offered another interpellative call to Americans to accept their collective identity and the accompanying necessary (if unfortunate) security precautions required for its preservation. Questions concerning the congruity of these measures with America's 'core spirit of liberty', in short, were already being foreclosed in advance.

In sum, this writing of post-9/11 America(ns) as undergoing a return to moderated normality represented a second attempt to write the War on Terror around a notion of temporal linearity. It did so by suggesting that although changes in security practice and infrastructure would necessarily accompany the experience or lessons learned on 11 September 2001, these changes were necessary to protect the continuity of something far more significant: the continuation of that society's core values and principles. And, as we have seen, not only could this construction of time only come into being at the exclusion of alternative writings of pasts, presents, and futures, it also offered a further valuable, a further critical, resource for the Bush administration's efforts to present this war and its components as a necessary, and legitimate, response to the (growing, not new) threat of terrorism. In short, what we have here is again neither an inevitable nor neutral description of time's passage and movements. But, rather, a(nother) political construction.

Writing progress into the War on Terror

The final construction we need to explore in this chapter differs a little from any of the five writings of temporality we have considered thus far. Where each of those was structured around a (specific, exclusionary, and political) attempt to position 9/11 as a bringer of *either* gradual transformation *or* qualitative rupture – whether in political violence, the American public, or in the security environment confronting America and its allies – this particular writing relates only to the subsequently instituted War on Terror itself. Specifically, I am interested here in the manifold, indeed near-ubiquitous, references to this conflict as a directional, progressive activity that would lead, ultimately, inevitably, to victory against terrorists and their unsavoury equivalents. These references, I argue, resonated entirely with the two writings we have explored in this chapter: conceptualising time once more, as a linear dynamic characterised by a pattern of continuous, evolutionary, change that could be at once known and foreseen. For, as we shall see, despite the persistent governmental efforts to frame the War on Terror as a radically new form of conflict lacking any directly predictable end point, the same representatives of the bush administration also worked hard to structure this war around claims to a privileged knowledge of that end point's final destination.

Consider, to begin with, Bush's (16/9/2001) remarks at the White House South Lawn that we began to explore in a different context within the previous chapter:

> My administration has a job to do, and we're going to do it. We will rid the world of the evil-doers. ... I can assure the American people I am determined, I'm not going to be distracted, I will keep my focus to make sure that not only are these brought to justice, but anybody who's been associated will be brought to justice. Those who harbor terrorists will be brought to justice. It is time for us to win the first war of the 21st century decisively, so that our children and our grandchildren can live peacefully into the 21st century.

We have, in this statement, an uncompromising rhetoric of sureness and predictability seeping through the President's every sentence. Substituting, here, a future of contingency, surprise, and unknowability, for one of foreseeable, knowable, indeed controllable, events, Bush is able to speak here with such confidence of events yet to occur. Guided by a resolute and determined administration, the future outcome of the first war of the twenty-first century appears now already certain for Bush – 'We will rid the world of the evil-doers.' And, what is more, the President is also already able to speak with such confidence on the pacific consequences of that forthcoming victory – 'so that our children and grandchildren can live peacefully into the 21st century'. Despite the new security challenges confronting Americans, their leaders, and allies, then, those same Americans could rest assured that justice awaited those who had so recently wronged them. Victory was, in this writing of temporality, simply assured. If not, entirely, predetermined, it was clearly certain.

As many of us will, no doubt, already know, the President returned to this image of a linear temporality culminating in inevitable triumph over the forces of terrorism with considerable persistence in the weeks that followed these early remarks. In his 29 September 2001 radio address to the nation, for example, he married this progressive dynamic to the discontinuous originary construction of 9/11 considered in the previous chapter, arguing: 'We did not seek this conflict, but we will win it. America will act deliberately and decisively, and the cause of freedom will prevail' (Bush, 29/9/2001). And, when questioned on the likely chronology of the newly launched Operation Enduring Freedom more specifically, he invoked it once more on 11 October, arguing:

> This particular battlefront will last as long as it takes to bring al Qaeda to justice. It may happen tomorrow; it may happen a month from now; it may take a year or two. But we will prevail. And what the American people need to know is what our allies know: I am determined to stay the course. And we must do so. We must do so. We must rid the world of terrorists so our children and grandchildren can grow up in freedom. It is essential. It is now our time to act. And I'm proud to lead a country that understands that.
>
> (Bush, 11/10/2001b)

As the third mensiversary of the 9/11 attacks rolled swiftly round, Bush again revisited this sense of unidirectionality – invoking, interestingly, a host of unnamed precedents for this present conflict's likely success. And, as in those considered directly above, he also drew here on a specific journey metaphor to capture this notion of linearity: a journey metaphor of which the President was rather fond in his rhetoric more broadly (Charteris-Black, 2005: 173):

> We still have far to go. And many dangers lie ahead. Yet, there can be no doubt how this conflict will end. Our enemies have made the mistake that America's enemies always make. They saw liberty and thought they saw weakness. And now, they see defeat.
>
> (Bush, 11/12/2001)

As the life of the unfolding War on Terror moved from weeks into years, the President had recourse to moderate this particular linear writing. Where his earliest pronouncements of future success had offered a *promise* of victory against American enemies, his subsequent employments of this construction suggested that this promise was already being actualised. Progress, we were told, was already underway: significant steps had already been taken along this path leading to inevitable triumph over terrorism. So, in a discussion of the nation's critical infrastructure the following year, for example, Bush (11/6/2002) cited the detainment of over 2000 terrorists – not suspected terrorists – as evidence of this ordered temporality, arguing:

> Interestingly enough, we've rounded up and detained over 2,400 terrorists, and that's good. It's not just us; our friends have, as well. We're making progress. You probably read in the newspaper, the number's now 2,401.

While his subsequent, 10 July 2002, remarks to a gathering of Federal employees located the agency underpinning these successes to date not only within the actions of his own administration but also, and equally, within a patriotic American public and their heroic troops defending their interests abroad. Although this journey was proving difficult, he told us now, neither the determination nor stamina of those charged with concluding it had waned in the ten months that had now passed since its inception:

> We're making progress overseas, we're making progress at home. We strengthened our aviation security and we're tightening our borders. We're stockpiling medicines to defend against bioterrorism. We're improving information-sharing amongst our intelligence agencies.

> We're taking new steps to protect critical infrastructure. But the important thing for the American people to know is that our public servants are working longer hours and working harder and working smarter to defend the American people. We will win the war on terror – no doubt in my mind – thanks to the heroism of our fighting troops, and thanks to the patriotism of our people, and thanks to the hard work of government officials here at home. There have been some extraordinary stories about some in your ranks – many in your ranks. And I want to share some of the stories with you so the American people fully understand the dedication of the people throughout our government.
>
> (Bush, 10/7/2002)

In August 2002, Bush returned to the earlier 'liberation' of Afghanistan from Taliban rule as a further key moment within this progressive dynamic, linking that liberation to the recently submitted Defense Appropriations Bill as an equally significant step along this unfurling march of History:

> And we're making pretty good progress. We're making pretty darned good progress. I laid out a doctrine that said, if you harbor a terrorist or you feed one of them, you're just as guilty as the terrorists, and the Taliban found out exactly what we meant. But I want – I want the youngsters here to understand that, when we went into that country, we went in as liberators, not as conquerors. We freed people from the clutches of a barbaric regime and, thanks to the United States and our friends and allies, many young girls now go to school for the first time in their lives. History will note that we didn't hit and run, that we stayed there. ... I think we've hauled in over 2000 of the enemy – 'we' being all kinds of people – the Philippines and Spain. Of course, the United States. We're making pretty good progress. We're getting them one by one. ... And we got a lot of work to do, we've got a lot of work to do. And that's why this budget I submitted is a significant budget. The House passed its version, the Senate passed its version. They've now got to get together as quickly as possible, as soon as possible, and get the defense appropriations bill to my desk nearly upon arrival. In other words, as soon as they get back from the recess, I need to sign the bill so we can plan for the war.
>
> (Bush, 15/8/2002)

A theme he revisited, finally, on 13 November 2002, when asked of the significance behind Bin Laden's continuing ability to evade capture:

> We're making great progress in the war on terror. Slowly, but surely, we are dismantling the terrorist network. We're finding their sanctuaries, we're holding people to account. Our coalition of freedom-loving nations is up to 90 now. There is an international manhunt on. I warn the American people that this is going to take time to achieve our objective. ... And we're not quitting. Slowly but surely, we're achieving our objective.
>
> (Bush, 13/11/2002)

If the President was wont to return to this writing of the War on Terror as a dynamic of evolutionary, gradual progress throughout its unfolding, he was far from alone in this within his administration (for further examples, see Bush, 1/10/2001, 4/10/2001a, 26/7/2002, 14/8/2002, 9/9/2002, 28/9/2002). The Vice President proved an equally persistent advocate of this writing in both its prospective and retrospective versions (see, for example, Cheney, 23/10/2001, 26/8/2002, 29/8/2002), arguing remarkably early into this war:

> [T]here is not a man or woman in this room who does not wait on the day that justice is delivered, as it will be ... terrorists and their supporters are, for the first time, worried about their own safety. They are being swept up in a worldwide roundup. Hundreds already have been arrested or detained. We have disrupted their operations and no doubt prevented some planned attacks. We're untangling the very complex financial network that funds terrorist activity. The terrorists have wealthy patrons. We are finding them and freezing their assets. So far, dozens of these individuals and groups have been identified. But there are others, and they too will be dealt with.
>
> (Cheney, 18/10/2001)

A theme he subsequently returned to when addressing his troops at Miramar Marine Air Station – 'The outcome of this conflict is certain, our nation has the strength and the patience required for a long struggle' (Cheney, 18/2/2002). And when reflecting on the import of the newly established Department of Homeland Security – 'To defeat them, we must be patient and resolute, and we must be organized for victory. The new department of homeland security will be crucial to our cause.

Whatever it takes abroad and at home, we will face this threat to America and we will prevail' (Cheney, 24/10/2002).

From the Department of Defense, similarly, Rumsfeld's 1 November 2001 news briefing warned against expectations of immediate, short-term successes in this conflict, while stressing at the same time the administration's resolve and determination to see it concluded:

> Victory will require that every element of American influence and power be engaged. Americans have seen tougher adversaries than this before, and they have had the staying power to defeat them. I think underestimating the American people is a big mistake. In the end, war is not about statistics, deadlines, short attention spans, or 24-hour news cycles. It is about will – the projection of will, the clear, unambiguous determination of the President and the American people to see this through to certain victory.
>
> (Rumsfeld, 1/11/2001)

Colin Powell (10/10/2001), in a briefing at the FBI Headquarters, stressed the role the state Department would play in progressing the War on Terror:

> Ladies and gentlemen, the campaign against international terrorism requires us to use every tool we have: political, diplomatic, intelligence, law enforcement, financial and military. Under the President's leadership, we are doing just that. Here today, diplomacy and law enforcement are standing side by side. United as a country, united as a people, we will be successful. As the President has said, we will prevail. And the State Department will play a role in this campaign and in this war, and we truly will win.

While the Homeland Security Director Tom Ridge (see also Ridge, 27/11/2001, 23/4/2002, 10/6/2002), finally, listed a plethora of developments that had been enacted to move this new war forwards towards victory:

> As a nation, we have learned a great deal about what it will take to secure our homeland in the six months since President Bush signed the Executive Order creating this office. The challenge is vast. But we are making real progress. ... We launched the largest criminal investigation in history. Passed the USA PATRIOT Act to bring federal law into the Information Age. Put hundreds of air marshals on our planes and

> new federal screeners in our airports. Deployed the National Guard to protect the border. Refocused the mission of the Coast Guard toward defending our coastline. Distributed more than a billion dollars to states' public health systems to build up their bioterror response. Acquired more than a billion doses of antibiotics. Stepped up security at our national monuments and nuclear power plants. Conducted a top-to-bottom security review of our entire energy infrastructure. Created the Homeland Security Advisory System, which has garnered widespread praise from first responders and law enforcement.
>
> (Ridge, 11/4/2002)

In short, when seen through this lens, military campaigns, detentions of suspects, the sharing of information, the freezing of assets, and even acquisitions of antibiotics all represented considerable, significant, steps along a pathway to triumph against terror. As victory had been promised – had been known in advance – by Bush and others in the immediate aftermath of 9/11, it was now already being delivered: was being delivered, indeed, so early into this new, multifaceted, conflict.

The politics of progress

It is, of course, tempting to approach the above statements of progress, direction, and future victory as mere rhetorical dressing for an administration embarking on war. Calls to arms and to battle right throughout history have been frequently couched in the language of impending, inevitable triumph, as we would no doubt expect (Graham et al., 2004). To view this particular language as only this, however, would be to miss the significance of this final writing for our exploration of time and politics in the War on Terror's unfolding. For, as the above passage hopefully indicates, this writing depended once more on a very particular vision of time: a vision of time as an ordered, broadly knowable, and singular dynamic. Written thus, although it may not (necessarily) have been possible to know the specific location of the War on Terror's conclusion, we could be certain of the general direction in which that conclusion lay. Even amid the unpredictability and surprise characterising this (potentially new) security environment, then, time, in this writing, possessed a discernible, directional, linearity. Under the guidance of a resolute and determined administration, future victory was certain. Under the guidance of a resolute and determined administration, future victory was already being brought into being.

Before turning to this writing's productivity for the Bush administration's new war, it will be valuable, briefly, to explore its structuration a little further. For, as with each of the accounts we have considered thus far, this unmitigated story of continual progress could only emerge through the exclusion of other plausible tales. And, in this case specifically, this writing relied quite clearly on the exclusion of any recognition of this war's failures and limits within its unfolding. While obstacles and difficulties had, of course, been encountered on the pathway to victory, these obstacles and difficulties were not to be interpreted as anything other than temporary setbacks. Their significance and status was, in this production, here minimal in comparison to this broader directional trend. To get a fuller sense of this exclusionary construction, consider, again, Bush's response to a question on Bin Laden's continuing ability to evade capture or death:

> We're making great progress in the war on terror. Slowly, but surely, we are dismantling the terrorist network. We're finding their sanctuaries, we're holding people to account. Our coalition of freedom-loving nations is up to 90 now. There is an international manhunt on. I warn the American people that this is going to take time to achieve our objective. ... And we'e not quitting. Slowly but surely, we're achieving our objective.
>
> (Bush, 13/11/2002)

As the President's remarks here neatly indicate, Bush is able to attribute a gradual, incremental directionality to the unfolding War on Terror only by offering a partial and selective reading of that war. In his response to the question, Bin Laden's continued existence is entirely absent, and what we are given, instead, is a relatively broad, relatively bland, account of the expanding coalition of freedom-loving nations, and the continued dismantling of 'the' (single, unitary) terrorist network. Presented in this way, the inability to capture Bin Laden, as Bush repeatedly reminded us (see, for example, Bush & Shinawatra, 14/12/2001; Bush & Franks, 28/12/2001), was not to be seen – could not be seen – as an indicator of failure within this unfolding war. For the parameters and mission of this conflict extended far beyond one man alone. As the 2006 White House report *9/11: Five Years Later: Successes and Challenges* made quite clear, this (new) War on Terror stretched quite some distance:

> Since the September 11 attacks, America and its allies are safer, but we are not yet safe. We have done much to degrade al-Qaida and its affiliates and to undercut the perceived legitimacy of terrorism. Our Muslim partners are speaking out against those who seek to use their

> religion to justify violence and a totalitarian vision of the world. We have significantly expanded our counterterrorism coalition, transforming old adversaries into new and vital partners in the War on Terror. In addition, we have transformed our governmental institutions and framework to wage a long-term struggle.
>
> (White House, 9/2006: 4)

As this report neatly demonstrates, this linear, progressive, writing of the War on Terror could only be sustained by (again) offering a violent and exclusionary – hence political – reading of this conflict. No failures (strategic, political, ethical, or otherwise) had here entered into this war. While impediments, pauses, or impermanent obstacles would no doubt present themselves in the journey towards conquest and peace, those impediments needed contextualising as but temporary and unfortunate departures from temporality's overarching, directional, march. As should be clear, we are here some distance, again, from any sense of time as a periodic, punctuated, unknowable process in this construction.

Finally, if this third narrative of temporal linearity only offered a partial reading of the Bush administration's new conflict, it also performed at least three significant discursive functions within this particular war. In the first instance, this construction of knowable futures worked to inscribe political and strategic clarity into the administration's actions at home and abroad. By positing and persistently returning to a singular, directional, temporal dynamic, the President and others were able to confidently assert, to confidently identify, any number of vindicating moments within their continuing confrontation with terror(ism). With the War on Terror's future outcomes here known and controlled by a determined and resolute administration, any sense of political impotence that had emerged in the wake of the 9/11 attacks was now, neatly, erased. If Bush, Rumsfeld, Powell, and others had been unable to foresee this new conflict's arrival, they could certainly – would certainly – contribute to its shaping and outcomes. In this sense, as noted in the previous chapter, this writing worked to reinscribe agency and confidence into the republican administration that had been so recently surprised, shocked, and traumatised by events far beyond their control.

Second, this production of a winnable war – or, later, of a war being won – also offered a powerful normative justification for this conflict's very existence. Tapping into just-war theory concerns relating to a conflict's probability of future success (Rengger, 2002: 358; Crawford, 2003: 7; Jackson, 2005: 137–9); the administration were able here to add further legitimacy to their new undertaking. If this confrontation had already been justified, first, by appeal to the agency of others in bringing it into being; second, by appeal to the continuing presence of imminent threat; and third, by appeal to the need for defending embedded civil liberties, this final construction of linear times offered another, still further, ethical justification for its very existence.

And, finally, as we have seen, this construction also offered yet another discursive resource for inscribing structural coherence into the War on Terror itself. As Tom Ridge's above remarks perhaps most vividly indicate, this imposition of a singular temporal directionality upon the War on Terror's discrete and multiple fronts served to link or equivalence any number of otherwise separable actions and policies under one (discernible, knowable) progressive dynamic. By subsuming the specific temporalities of military operations, homeland security measures, and so forth under this one coherent historical course, this writing made possible the aggregation of otherwise distinct strategic manoeuvres into a unitary and coherent discursive formation. Seen through this lens, the War on Terror once more could be approached not as a partial and incomplete imposition against an undecidable ontological horizon, but, rather, as a fully formed, consistent, albeit multifaceted, response to an equally unitary and fully formed enemy other.

In sum, the Bush administration's persistent writing of the War on Terror as a continued, progressive development again represented neither an inevitable nor a neutral reflection of an underlying empirical reality. Rather, this writing offered a contingent and exclusionary – and therefore, again, political – reading of this war: a reading with discernible implications for the war itself.

Conclusion

In this chapter we have seen that representations of discontinuity were far from alone in the Bush administration's writing of their War on Terror around claims to temporality. Each of the above constructions – of 9/11 as precedented continuity, of a post-9/11 return to moderated normality, and of the War on Terror as a progressive dynamic – relied on a thoroughly distinct sense of time to any of those in Chapter 3. Where notions of discontinuity endow temporality with properties of qualitative transformation and discrete periodicity, these three accounts presented time instead as a directional, linear, and ordered dynamic. If the War on Terror was still characterised or coloured by change, in these writings, those changes were emphatically not unpredictable, contingent, or immeasurable.

Once we begin to take these representations of temporal linearity quite seriously we are in a position to garner a further understanding of both the War on Terror, on the one hand, and time, on the other, than that with which we were left at the conclusion of Chapter 3. With regard to the former, we have seen that this conflict was rendered possible, necessary, and legitimate not only within the claims to uniqueness and exceptionality that have captured the attention of critical scholars to date. But also, and equally, within claims to continuity and progression. And, grasping this moment of discursive heterogeneity is important, of course, not only to achieve a fuller account of the War on Terror's unfolding on its own terms – a worthwhile pursuit in itself – but also because it allows us to glimpse this conflict's undecidable, and incomplete, existence to which we pointed in Chapter 1. That 9/11, America(ns), the terrorist threat, the War on Terror itself, and so forth could be positioned – persuasively and frequently simultaneously – into such radically distinct temporal shapes indicates how illusory it is to approach any of these entities as fully formed, extra-discursive, objects of knowledge. Their meaning and significance was the product of human endeavour and decision.

With regard to the latter, we now have still further support for our claim to the status of time as an inherently, necessarily, political phenomenon. For, not only did these three productions of temporal linearity again perform key constitutive functions in this particular context, notably:

- Problematising the contemporary terrorist threat within a specific historical trajectory of increasingly destructive assaults and attacks
- Legitimising the institution of the War on Terror within a reading of historical violences unchecked
- Justifying the institution of potentially controversial moments within this conflict by disavowing the exceptionality or uniqueness of those responses
- Writing political certainty into the (likely) future outcomes of the War on Terror
- Introducing structural coherence into this war by linking its disparate moments into a singular historical pathway towards victory

They also, again, militated against alternative understandings of that conflict's status, import, and historical position. And, as we shall see now in Chapter 5, this is not, yet, all we can say on either of these themes.

5 Writing Timelessness

In the preceding chapters we encountered the emergence of two very different, yet similarly prominent, conceptions of temporality within the War on Terror. On reflection, it seems almost remarkable now that this one discursive framework could accommodate such radically distinct claims to temporal discontinuity *and* linearity: not least given the speed with which each of these shapes surfaced following the events of 11 September 2001. Discussions of rupture and revolution, as we have seen, coexisted surprisingly neatly with statements of historical continuity or evolution. No overt sense of contradiction, ambiguity, or incoherence appeared to accompany the administration's mix-and-match approach to the temporal in their discussing this conflict.

Perhaps even more remarkable still, however, is the ease with which the same objects of discourse were concurrently positioned within these heterogeneous conceptions of time. 9/11, the American public, the threat of terrorism, and, indeed, the War on Terror itself were all inserted into these two discrete shapes by the very same representatives of the Bush administration, by the very same representatives speaking, no less, at the same historical juncture: informing onlookers, on the one hand, that nearly everything of import had changed so suddenly, so dramatically, on that infamous, ill-fated, morning. And on the other, encouraging them to share in historical precedents; to participate in the bringing about of (knowable) futures to come; and to accept minor (security) changes as a means of protecting the continuation of core rights, values, and norms.

To make matters even more complicated and interesting now, this chapter introduces yet another temporal shape that rose to prominence in the administration's writing of this unfolding war. This final understanding turned emphatically away from any sense of time as a site of dynamism and change: substituting claims to qualitative *and* quantitative chronological transformation with a story, instead, of perpetual return. For, these writings of timelessness produced temporality as a site of recurrence and sameness: a universal, unchanging, and static horizon comprising equivalent, synonymous, pasts, presents, and futures. Understood in this way, the War on Terror's events, actors, and status were all reduced, effectively, to reincarnations of earlier historical dynamics. The distance of time had brought nothing new – nothing different – to its observers and subjects.

The discussion that follows centres on two distinct, yet related, uses of this third temporal shape by the Bush administration. The first of these concerns the persistent effort to construct this new conflict as the most recent recycling of a permanent battle between good and evil. The second, it's writing as the latest moment in a similarly enduring struggle between freedom and fear. Positioned as neither departure nor development, these writings together, I will argue, worked to insert the War on Terror into a perpetual antagonistic confrontation between two essentialised entities. And, as we shall see, they also were integral in offering this conflict further legitimacy, coherence, and political necessity. Let us turn to each now in turn.

Writing good versus evil

A good deal has already been written on the Bush administration's penchant for explicitly moral rhetoric in their writing of the War on Terror. For some, the apparent appeal of this language is traceable to the 'Manichean psychic template' of an American population with a rather unusual proclivity for religion and churchgoing: unusual, at least, in the contemporary Western world (Tracy, 2005: 94; see also Singer, 2004: 4–5; Jackson, 2005: 122;). For others, its attractiveness may be traced more specifically to a distinctively republican form of conservative politics that continues to exist within – if not precisely throughout – the United States (compare Lakoff, 2001: 5; Edwards, 2004: 164). My own interest in the administration's repeated employment of this language, however, relates more specifically to the ways in which their persistent return to essentialist, metaphysical, notions of good and evil positioned the War on Terror within a very specific notion of temporality. Or, perhaps better, to the ways in which this language worked to *exclude* any meaningful consideration of temporality as a site or process of change from this particular war. For, as we shall see in the discussion that follows, absolutist invocations of morality not only erase the relevance of times past in the form of political amnesia noted by Campbell (2002: 14).

They also coerce against any understanding of the specificity of times present and future. In order to clarify and elaborate on this claim, allow me to briefly return first to an earlier discussion.

In Chapter 3, we encountered the Bush administration's frequent scripting of 9/11 as a moment of normative rupture. A key part of that writing, I argued, related to the depoliticised and rather simplistic framing of those events as an evil act conducted by (similarly) evil perpetrators. As the President's remarks of 13 September 2001 illustrate quite succinctly, this twin writing of evil offered a particularly powerful discursive resource for a President preparing for war:

> Civilized people around the world denounce the evildoers who devised and executed these terrible attacks. Justice demands that those who helped or harbored the terrorists be punished – and punished severely. The enormity of their evil demands it. We will use all the resources of the United States and our cooperating friends and allies to pursue those responsible for this evil, until justice is done.
>
> (Bush, 13/9/2001b)

Bush's remarks here position 9/11 as a shattering of an antecedent, stable, horizon of morally acceptable actions and behaviours. In so doing, he invokes a distinctive sense of temporal discontinuity by stressing the enormity – the exceptionality – of the evil that had broken a hitherto normal existence that morning. This much we know, already, from our earlier discussion. What is interesting, however, is that upon turning the clock forwards just slightly, we encounter a similar, yet distinct, employment of this metaphysical language. Not, here, in reference to the 9/11 attacks, but, rather, in reference to the subsequent War on Terror itself. And, as the President's 4 October remarks on humanitarian aid to Afghanistan vividly suggest, this important discursive shift offered a rather different sense of this conflict's temporal location to any of the framings we have been exploring thus far:

> We've made it clear, this is not a war between Christianity or Judaism and Islam. As a matter of fact, the teachings of Islam make it clear that peace is important, that compassion is a part of life. This is a war between good and evil. And we have made it clear to the world that we will stand strong on the side of good, and we expect other nations to join us. This is not a war between our world and their world. It is a war to save the world. And people now understand that.
>
> (Bush, 4/10/2001b)

Bush's immediate concern in this passage is to distance the War on Terror from any efforts to understand it as a clash of civilisations. As we were then repeatedly informed (see, for example, Bush, 17/9/2001b, 19/9/2001, 20/9/2001, 27/9/2001, 10/10/2001b), this conflict could not be – must not be – perceived as a war against Islam. Whatever the plausibility of this reading, our own interest in it relates here to the way in which this spatial unbounding of the War on Terror also contains within it a temporal unbounding. For, by writing it, simply, as an extra-territorial war to save the world as we know it, Bush simultaneously detaches this conflict from any specific, historical grievances or causes. The particular temporal coordinates of its occurrence are here irrelevant for its understanding: this conflict could equally have occurred at any other historical juncture. What we are confronting here then, in short, is a confrontation between two antagonistic moral forces: those charged with protecting (an imaginary, unitary) normative order and those that had demonstrated their determination to destroy it.

As is already well known, the President returned to this particular framing of the War on Terror as a clash between good and evil with considerable gusto throughout the months that followed these remarks (see, for example, Bush, 13/10/2001, 29/11/2001, 6/7/2002). In his 20 November appearance with President Arroyo of the Philippines, for example, Bush defended the White House's continued closure to the public by citing the omnipresence of evil:

> I know a lot of Americans look forward to touring the White House during this period of time. But we're in extraordinary times, and as I said yesterday, evil knows no holiday; evil doesn't welcome Thanksgiving or Christmas season. And in these extraordinary times we're taking extraordinary measures. It is further reason why we must continue to wage a diligent and consistent fight against terror, and to rid the world of terror, to make our country safe so we can have tours at the White House.
>
> (Bush & Arroyo, 20/11/2001)

A theme he subsequently revisited in justifying his administration's decision to expand the funding available for preventing bioterrorism:

> History has called us into action, here at home and internationally. We've been given a chance to lead, and we're going to seize the moment in this country. As we've mentioned more than once, what we do here at home is going to have lasting impact for a long time.

> And I want to tell you what we're doing abroad is going to have lasting impact, as well. I view this as an opportunity to secure the peace for a long time coming. I view this as a struggle of tyranny versus freedom, of evil versus good. And there's no in between, as far as I'm concerned. Either you're with us, or you're against us.
>
> (Bush, 5/2/2002)

The Vice President, similarly, appeared keen to view the War on Terror as a conflict against a single, undifferentiated, evil entirely incompatible with the views held by those (sincere) advocates of Islam:

> A few young men may be deluded into dying for the terrorist vision of the world, but only the terrorists themselves want to live in such a world, where law is replaced by brute force and morality is defined by vicious and violent men. The world understands that we do not fight a religion. Ours is not a campaign against the Muslim faith. Indeed, the innocent victims of these terrorists include many Muslims. This is a struggle against evil. That is why people in every part of the world and of all faiths stand together against this foe.
>
> (Cheney, 18/10/2001)

As did Colin Powell (19/11/2001) in his remarks in Kentucky, within which he argued: 'these criminals have no religion, and they have no human cause. Their goal, and the goal of all like them, is to divide and embitter people. They are evil merchants of death and destruction.' From the (then) Office of Homeland Security, Tom Ridge personified this confrontation in the figure of Bin Laden himself, describing a recently released recording of his as: 'from an evil heart, a hateful heart and an evil mind and an evil man' (cited in Kozaryn, 18/11/2002). While Donald Rumsfeld (18/12/2001), finally, and unusually, rendered the subjective, constructed, character of this evil absolutely explicit in his 18 December 2001 claim that 'There's no question but that the coalition forces across the globe that have decided that terrorism is bad and evil and dangerous and has the potential of killing tens of thousands of human beings – innocent human beings.'

This persistent return to a Manichean moralised language represented a key contributory moment in the War on Terror's construction around a specific conception of time as a recurrent, cyclical, process. In the first instance, this powerful and provocative lexicon worked to position 9/11 and the continuing terrorist threat as an absolute moral transgression entirely removed from any historical problematic of origins, factors, or causes. With goodness and evil here dissociated from any form of contextual or discursive framework, this writing reduced the War on Terror to the latest moment or turn, simply, in a permanent, enduring, battle between good and evil. For, as Bush (30/3/2002) argued in a radio address to the nation in the following March:

> On September the 11th, a terrible evil was done to our country. America and the civilized world are now joined together in a great struggle against enemies who have no regard for innocent life. Grave challenges and dangers face us in this war. We cannot predict every turn ahead of us. Yet in this season, we are assured that history is of moral design. Justice and cruelty have always been at war, and God is not neutral between them.

In the second instance, this writing also worked to deny any meaningful historicity from the *evaluation or judgement* of the War on Terror's normative status. Not only was this war here reduced to a simplistic, eternal, Manichean struggle devoid of politics or history proper, its morality or legitimacy could be assessed by similarly permanent, similarly enduring, normative standards. As Bush (11/10/2001a, 11/10/2001b) informed us on several occasions, to understand this war it was enough, simply, 'to know that evil like goodness exists'. For, although, 'it may hurt some people's feelings, it may not be what they call, diplomatically correct ... I'm calling evil for what it is. Evil is evil, and we will fight it with all our might' (Bush, 11/6/2002). And, as his remarks of February 2002 neatly suggest, Bush's own moral clarity concerning this battle's legitimacy was founded on grounds some distance removed from Rumsfeld's more subjective reflections noted above:

> [F]aith shows us the reality of good, and the reality of evil. Some acts and choices in this world have eternal consequences. It is always, and everywhere, wrong to target and kill the innocent. It is always, and everywhere, wrong to be cruel and hateful, to enslave and oppress. It is always, and everywhere, right to be kind and just, to protect the lives of others, and to lay down your life for a friend.
>
> (Bush, 7/2/2002)

The implicit, dual, construction of timelessness underpinning this writing of the War on Terror into a perpetual moral conflict was also, importantly, supplemented by more explicit reflections on this war's direct historical equivalents. The earliest such analogy, of course, and that which was thereafter most frequently returned to, emerged within comparisons of this conflict to the earlier fight against the forces of fascism and National Socialism within World War II (Noon, 2004; Jackson, 2005: 41–4). As the President noted in his 11 October tribute at the Pentagon Memorial, for example:

> In New York, the terrorists chose as their target a symbol of America's freedom and confidence. Here, they struck a symbol of our strength in the world. And the attack on the Pentagon, on that day, was more symbolic than they knew. It was on another September 11th – September 11th, 1941 – that construction on this building first began. America was just then awakening to another menace: The Nazi terror in Europe. And on that very night, President Franklin Roosevelt spoke to the nation. The danger, he warned, has long ceased to be a mere possibility. The danger is here now. Not only from a military enemy, but from an enemy of all law, all liberty, all morality, all religion. For us too, in the year 2001, an enemy has emerged that rejects every limit of law, morality, and religion. The terrorists have no true home in any country, or culture, or faith. They dwell in dark corners of earth. And there, we will find them.
>
> (Bush, 11/10/2001a)

The President's remarks here produce an explicit historical equivalence between the new War on Terror and World War II (see also, Bush, 17/4/2002, 30/4/2002, 9/5/2002, 23/5/2002, 6/12/2002, 9/7/2005). As Roosevelt and America had awoken to the menace of Nazism in 1941, Bush and (the same) America had awoken to the same threat of evil in 2001. Although separated by a distance of some 60 years, the rejection of faith, morality, and law shared by each of these enemies rendered the two essentially identical. To understand the status or significance of either of these conflicts, it was simply unnecessary to reflect on when each had occurred. For, temporality as a genuine dynamic or process of change here bore no relevance for knowing this war or its outcomes. An acquaintance with this cyclical, Manichean, history possessed, for the President, far greater utility than did any knowledge of the interval separating these events.

Although the Pentagon Remembrance Service one month after 9/11 perhaps provided an understandable location for exploring the War on Terror's historical equivalents, Bush's administration returned to the experiences of the World War II with considerable regularity throughout the following months. Rice's (10/4/2002) remarks at the National

Commemoration of the Days of Remembrance, for example, explicitly paralleled the Holocaust's evil with that now confronting America while making clear, in the process, the universality upon which these evils could be condemned:

> We gather today to remember that evil is real and present in our world. We gather to remember that hatred and bigotry are always and everywhere wrong. ... Last year, when the President spoke here, the Holocaust seemed somewhat removed from our era – part of a bloody century now behind us. Sadly, this year we need no prompting to appreciate the Holocaust's importance and its relevance. Fanatical, unreasoning hatred has intruded upon our lives in ways that no one could have imagined months ago. ... And I have thought about that couple from the ghetto even more in the days since September 11. Because right now, all of us are enduring a time of testing, loss, and fear; a time when our vulnerability to evil and the certainty of our mortality are all too clear; a time when once again our intellect is insufficient to answer the question, 'Why?' And at these times more than ever, we are reminded that it is a privilege to struggle for good against evil.

While, John Ashcroft (7/8/2002), similarly, chose to revisit the Nuremburg Trials in outlining the need to confront this new force for evil:

> Throughout our history, there have been periods – such as now – when the most essential security of our nation is threatened. ... At the conclusion of World War II came the reckoning at Nuremberg. Supreme Court Justice Robert Jackson led the prosecution of 21 Nazi defendants for crimes against their countrymen, crimes against their neighbors, and crimes against humanity. All pleaded not guilty. Some claimed that they were merely following orders. Others disputed the jurisdiction of the court. But Jackson successfully argued their guilt with a sense of urgency borne of a civilization threatened by a new force for evil. 'The wrongs which we seek to condemn and punish have been so calculated, so malignant and so devastating,' said Jackson, 'that civilization cannot tolerate their being ignored because it cannot survive their being repeated.' It is now as it was then. A calculated, malignant and devastating evil has arisen in our world. Civilization cannot ignore the wrongs that have been done. America will not tolerate their being repeated. Justice has a new mission, a new calling against an old evil.

With good and evil *once more* here confronting each other in an absolutist struggle of epic proportions, indecision and inaction simply could not now be countenanced. As Ashcroft makes abundantly clear, the certainty of our moral convictions and actions in 2002 needed to parallel that of Supreme Court Justice Jackson more than 50 years prior.

Although a powerful analogy in itself, this writing of World War II and the War on Terror as equivalent moments in a permanent Manichean conflict was also supplemented by invocations of other earlier battles against evil. Whether by reference to the similarly totalitarian threats of the Soviet Union or Pol Pot's Cambodia (see Bush, 20/9/2001, 23/5/2002, 23/11/2002, 11/6/2004; Rice, 8/4/2002; Cheney, 9/6/2004). Or in the parallel with the threat of organised crime confronted by America in the early twentieth century (Ashcroft, 25/10/2001). The Bush administration repeatedly inserted this unfolding conflict into a (glorious) American history of fighting immorality. In this sense, as suggested above, the 'first war of the twenty-first century' was increasingly coherently positioned as a conflict not (only) against a radically new form of enemy, nor (even) against an increasingly sophisticated existent threat, but rather, and far more simply, as a war against 'history's latest gang of fanatics trying to murder their way to power' (Bush, 11/9/2002b). As the President suggested on the third anniversary of the 9/11 attacks: 'the struggle of good against evil was compressed into a single morning' (Bush, 11/9/2004).

The politics of good and evil

This writing of the War on Terror into a perpetual struggle against evil offered a markedly distinct understanding of this political discourse to any of those we have considered in the two previous chapters. Offering almost no space at all for acknowledging any of this conflict's specificities, this construction simply inserted the Bush administration's new war into a recurrent, timeless, battle for good. With temporality reduced, now, to an eternal system of sameness or equivalence, nothing, really, appeared to have changed after all. Evil had, simply, always existed; fanatics had always tried to murder their way to power; and the (American) opponents of evil had always confronted these pernicious certainties. From such a perspective, the attacks of 9/11 and the very need for any form of war against terrorism here served not as a *revelation* of evil's presence in the world, but, rather, as the *latest reminder* of this unfortunate reality. In a very tangible and interesting way, then, this writing of the War on Terror as a confrontation of good versus evil could only exist through the exclusion of any meaningful notion of historical change. This conflict was reduced, blandly, to a detemporalised, Gnostic, Manichean confrontation: a Manichean confrontation assessable by similarly detemporalised perpetual truths. As America(ns) (and certain allies) had arrested the spread of evil at earlier critical junctures, so they must – indeed would – arrest the spread of evil today.

A second, more obvious, exclusion of this writing of timelessness concerned the absence in this story of any number of potentially relevant evils that have passed by unchecked within the American socio-political experience. Without, again, wishing to engage in any moral accounting, an alternative writing of America more carefully attuned to the history of slavery (see, for example, Thomas, 1996), the conquest of native Americans (see, for example, Campbell, 1998b), or, indeed, the range of contemporary military interventions undertaken by its political leaders (Jackson, 2005: 78–9), would no doubt have offered a less vainglorious reading of the state than that permitted in this writing. For, as Rice's (10/4/2002) comparison of the War on Terror to the Holocaust explicitly noted, an awareness of any number of pauses, intervals, or failures inside this history could quite easily have undermined this construction of an eternal, unrelenting, battle against evil:

> We gather to remember that the commission of monstrous sin requires not our consent, but only our indifference, our neutrality, or our silence. We gather to light six candles, so that we may never forget six million acts of murder.

As Rice's remarks here suggest, where this cyclical writing of the War on Terror offered a specific and exclusionary – hence political – understanding of the (non-)passage of temporality, it also offered a specific and (potentially) exclusionary reading of American history as an unmitigated battle for good.

If this Manichean narrative did not offer an inevitable, natural, or uncontestable reflection of the War on Terror's historical location, neither, of course, did it represent a neutral presentation of this unfolding conflict. In the first instance, this writing quite clearly functioned as a powerful discursive strategy for (re)inscribing significance into this conflict's conduct and outcomes. Positioning this war as important not because of the need to re-establish American (or international) security in the wake of an (newly revealed) exceptional threat. Nor, because our earlier failures at confronting terrorism had rendered this enemy even more determined – even more deadly. This writing rendered the War on Terror significant, instead, *because of* its location in this perpetual battle against evil. By producing this conflict as something far greater than a mere struggle for safety, the Bush administration were able to construct their new war as a necessary contributory moment in a monumental (a)historical struggle for good. Written in this way, any refusal to participate in, contribute to, or even, simply, accept this conflict's necessity risked bringing with it an impression of complicity with evil. For, with the world produced in such stark, binary, language, any and all efforts at dissent or resistance had been already condemned before their emergence.

Second, this Manichean approach to temporality as a static, recurrent, unchanging horizon also assisted the administration's efforts to position the War on Terror as a morally legitimate conflict. Although, constructed upon rather different grounds to the Just War-inflected discussions we have considered thus far, this writing offered powerful deontological *and* consequentialist legitimisations for the Bush administration's launch and continuation of this war. Positioning it, now, within this permanent struggle with evil, this writing not only rendered inaction an inconceivable abdication of an absolute, detemporalised, moral imperative. Its structuration around the host of historical analogies considered above also drew our attention to the unthinkable consequences that would no doubt follow any quiescence or complicity with the contemporary terrorist threat. In this sense, the fight against evil needed to occur not only because of its status as a morally legitimate – a morally just – course of action. But also because this struggle itself would engender desirable outcomes stretching far beyond American borders. As Bush (5/9/2002) argued in the approach to the first anniversary of the 9/11 attacks:

> If we stay steadfast and strong, if we're willing to delineate between good and evil, if we're willing to continue to work with others to protect freedom, we can achieve peace. We can achieve peace not only for our own citizens, but since we value all human life and everybody matters, we can achieve peace in parts of the world that have quit on peace.

While his subsequent remarks at the 2005 commemoration of Veterans Day rendered the unthinkable consequences of doing nothing explicit:

> And the civilized world knows very well that other fanatics in history, from Hitler to Stalin to Pol Pot, consumed whole nations in war

> and genocide before leaving the stage of history. Evil men, obsessed with ambition and unburdened by conscience, must be taken very seriously – and we must stop them before their crimes can multiply.
> (Bush, 11/11/2005)

In short, where the final linear writing of temporality that we considered in our previous chapter offered a powerful discursive resource for inscribing political certainty into this unfolding conflict, this cyclical writing of temporality offered this conflict an equally significant feeling of normative certainty.

Third, as a number of authors have already noted, this production of the War on Terror as a repetition of earlier struggles with evil also offered an important discursive resource for justifying the institution of unprecedented, and potentially controversial, measures in its conduct. At the signing of the 2001 USA PATRIOT Act, for example, Bush (26/10/2001) suggested, 'This legislation is essential not only to pursuing and punishing terrorists, but also preventing more atrocities in the hands of the evil ones,' a theme returned to in Ashcroft's (7/11/2003) subsequent, and more explicitly cyclical, framing of this Act's import:

> Today, in the morning of a new millennium, we can look back on almost a century of unfathomable evil. Voices of hate rose to power in great nations and millions were murdered in the name of totalitarian creeds – communism, fascism, National Socialism. We look back and grieve. And with the hindsight of history, we ask ourselves, 'What if one more voice had spoken out against hate and taken a stand against evil? Could it have made the difference?' We cannot pierce time to save the innocent. We can only look to the future and vow, 'Never again.' ... Following September 11th, Democrats and Republicans in Congress united and passed the Patriot Act with wide, bipartisan support – a law that provides our nation with improved tools to fulfil the first responsibility of government: to preserve the life and liberties of the people.

The establishment of military tribunals for trying enemy combatants were also, similarly, justified by their location in this recurrent Manichean history:

> [T]o set up the option to use a military tribunal in the time of war makes a lot of sense. We're fighting a war, Terry, against the most evil kinds of people. And I need to have that extraordinary option at my

> fingertips. I ought to be able to have that option available should we ever bring one of these al Qaeda members in alive. It's our national interests, it's our national security interests we have a military tribunal available. It is in the interests of the safety of potential jurors that we have a military tribunal. These are extraordinary times. And I would remind those who don't understand the decision I made that Franklin Roosevelt made the same decision in World War II. Those were extraordinary times, as well.
>
> (Bush, 19/11/2001)

As, more recently, were the efforts at reconstructing Iraq, as the President reflected on his second term in office:

> That's what the Iraqis are doing. They're coming up with a doctrine that will survive the years so that self-government and freedom prevail. And then they'll be voting on the document in October. And then they'll elect a permanent government in November. Democracy is moving forward, and that's part of laying the foundation for peace. We have done this type of work before in our nation. We have fought evil before. We have been through ideological struggles. Your dads and granddads fought against the Nazis and fought against the Japanese. It was an ideological struggle against an enemy that was ruthless. And we prevailed. We prevailed in more ways than one. We prevailed militarily, but we also helped spread democracy. We laid that foundation for peace for the next generation coming up.
>
> (Bush, 3/8/2005)

Whatever we may think of any of these developments, the point here is that their writing into a perpetual Gnostic battle added greatly to their sense of ethical legitimacy. Ruthlessness had always been combated with extraordinary measures: nothing new, nothing novel, characterised this latest such effort.

Fourth, as the above examples explicitly suggest, this writing of timelessness also functioned to inscribe further structural coherence across the War on Terror's manifold developments. By locating the USA PATRIOT Act, military tribunals, and the reconstruction of Iraq, for example, as equivalent events within a perpetual battle for good, this narrative positioned these potentially separable developments once more as inherently, self-evidently, connected. By subsuming these occurrences under a single discursive grammar, the administration was again able to present their War on Terror as a meaningful structural totality, locatable within a singular detemporalised history. And, finally, although less explicit than the writing we will turn to next, this construction of the War on Terror also contributed to the positioning of its parameters and import far beyond the US borders. By depicting the War on Terror as possessing a (moral) significance far transcending any search for security, this writing helped Bush and his staff to present this (latest) war as a truly global concern. For, once we recognised that goodness, of course, is not limited to America(ns), this conflict emerged here as quite far removed from narrow realpolitik and the worldliness of parochial national interests.

To summarise briefly, this first writing of the War on Terror around a notion of timelessness offered a radically incommensurate conception of temporality to either of those considered in our preceding chapters. It did so, of course, by conceptualising history neither as a process of punctuated periodicity nor as a dynamic of incremental, evolutionary development, but, rather, as a perpetual, continual struggle between – here – the forces of good and the forces of evil. And, as we have seen, this writing must again be approached as an inherently political construction: first in its excluding alternative readings of time and the War on Terror itself; and, second, in its more constitutive role in helping to bring this particular discourse into being as both a necessary *and* morally legitimate enterprise.

Writing freedom versus fear

The second discernible construction of timelessness that emerged in the aftermath of the 9/11 attacks concerned the Bush administration's persistent attempt to insert the War on Terror into a similarly enduring, perpetual conflict: an enduring, perpetual conflict not, here, between goodness and evil, but, rather, between freedom and fear. As several authors have argued (Pieterse, 2004: 120; Kennedy & Lucas, 2005: 325; Lewis, 2005: 118; Thomassen, 2005: 295), the concept of freedom performed a central and prominent signifying role in this new war from its very inception: not least, of course, in its emergence as a key referent object of the 9/11 attacks and a signifier of two subsequent military operations in Afghanistan and Iraq, to give but two prominent examples.

However ambiguous or hypocritical (compare Van Alstyne, 2002: 79; Der Derian, 2003: 20–2; Schehr, 2006: 143–4) the administration's usage of this language, my interest in it here relates to its role in the positioning of the War on Terror as the latest moment of another permanent, transcendental, battle. As with the good-versus-evil writing considered above, it is important that we see this positioning as again mobilising an inherently cyclical notion of temporality: relying, as it did, on an understanding of time as a perpetual process of repetition and recurrence. In this sense, as we shall see, this construction once more excluded any meaningful notion of time as a site of progression, dynamism, or change from the unfolding War on Terror: problematising still further those claims to both discontinuity and linearity considered in Chapters 3 and 4.

To begin this final substantive section of our discussion, consider the following statement from Bush's (20/9/2001) early address to Congress that has already emerged in our exploration of time and the political:

> The advance of human freedom – the great achievement of our time, and the great hope of every time – now depends on us. Our nation – this generation – will lift a dark threat of violence from our people and our future. I will not relent in waging this struggle for freedom and security for the American people; freedom and fear, justice and cruelty have always been at war and we know that god is not neutral between them.

Bush's remarks here represent some of the most powerful and evocative of any he offered throughout the War on Terror's construction. Metaphor – 'a dark threat of violence'; interpellation – 'our nation – this generation'; and divine inspiration – 'we know that god is not neutral', all ring loudly in his words, affording his statement of political purpose a sense of poetry and colour. What is more interesting for us, however, is the way in which this latest framing of the War on Terror operates through a similarly dual disavowal of temporality's significance to the writing of good and evil we have just been exploring. In the first instance, as with that other dichotomy, this framing of war works once again to isolate the unfolding conflict from any meaningful historical problematic. Causation, justification, and politics are all (now, again) rendered redundant for our understanding this war: we are simply encountering a reincarnated, eternal, and essentialised conflict between freedom and fear, justice and cruelty. Equally important, however, is the way in which this framing also functions to deny the notion of freedom itself any chronology or historicity. For, as Bush (29/1/2002) argued in his subsequent State of the Union address: 'America will lead by defending liberty and justice because they are right and true and unchanging for all people everywhere.' Conceptualised thus, the War on Terror is effectively reduced here to a war without a history for a concept without a history: ruptures, punctuations, linearities, and trends all similarly subsumed under a perpetual, cyclical, truth. And, where Bush concludes his remarks by invoking God's partiality in this second Manichean battle, the value of this writing for inscribing normative, indeed moral, significance into this war is already explicit. The War on Terror must be fought (and will be won), because the struggle for freedom must continue (and will continue) under the guidance of the American President.

This twin effort at detemporalisation was again returned to persistently by the Bush administration throughout the earliest months of the War on Terror. An early example of this writing was found, for example, in Bush's (6/11/2001) November remarks at the Warsaw Conference for Combating Terrorism:

> I thank all the nations of Central and Eastern Europe at this conference. You are our partners in the fight against terrorism, and we share an important moment in history. For more than 50 years, the peoples of your region suffered under repressive ideologies that tried to trample human dignity. Today, our freedom is threatened once again. Like the fascists and totalitarians before them, these terrorists – al Qaeda, the Taliban regime that supports them, and other terror groups across our world – try to impose their radical views through threats and violence. We see the same intolerance of dissent; the same mad, global ambitions; the same brutal determination to control every life and all of life.

Drawing on the Cold War historical analogy that we encountered in the first above discussion of timelessness, this writing once again positions the fight against terrorism within a perpetual battle being fought by the defenders of freedom and dignity. The contemporary threat posed by al-Qaeda, the Taliban and others, in this writing, is emphatically not the unprecedented or historically unique global entity it was in the discontinuous writings discussed in Chapter 3. Instead, it is, effectively, a re-emergence of the same threat as that which was posed by the fascists and totalitarians before. And, by moving seamlessly between a construction of the Central and Eastern European nations to whom he was speaking as historically other ('for more than 50 years, the people of *your* region suffered'), to an interpellative hailing of the same nations within a contemporary intercontinental collective identity ('Today *our* freedom is threatened once again'), this statement demonstrates the power of this writing for the (re)positioning of this war as a truly global concern. For, as Bush's subsequent (25 March 2002 and 17 April 2002) return to this usage of historical analogy suggested, this construction had considerable capacity to incorporate former allies *and* enemies alike within his administration's newest battle for freedom:

> The spirit of liberty has always had enemies. It's important for us to remember our history, that there are some who can't stand liberty and freedom, as we learned so tragically on September the 11th. The terrorists who attack us hate the very idea of human freedom. ... One ancient Greek statesman put it this way. He said, 'We are free to live exactly as we please, yet we are ready to face any danger.' He might as well have been talking about the United States of America. Fortunately, we don't face danger alone. There are other people who love freedom, as well, in this world. Greece and America have been firm allies in the great struggles for liberty. Greece was one of only a handful of countries that fought alongside the United States in every major 20th century war. And Americans will always remember Greek heroism and Greek sacrifice for the sake of freedom. As the 21st century dawns, Greece and America are – once again stand united; this time in the fight against terrorism. Greece is a part of the coalition of nations helping bring justice to those who would harm us, harm the people of Greece, harm anybody who love freedom. The United States deeply appreciates the role Greece is playing in the war against terror. We thank Greece for access to the military base in Souda Bay, Crete, and for providing a naval frigate for patrolling the Arabian Sea.
>
> (Bush, 25/3/2002)

> And these hopes are universal in every country and in every country – in every culture. Men and women everywhere want to live in dignity to create and build and own, to raise their children in peace and security. The way to a peaceful future can be found in the non-negotiable demands of human dignity. Dignity requires the rule of law, limits on the power of the state, respect for women, private property, equal justice, religious tolerance. No nation owns these principles. No nation is exempt from them. Sixty years ago, few would have predicted the triumph of these values in Germany and Japan. Fifteen years ago, few would have predicted the advance of these values in Russia. Yet, Americans are not surprised. We know that the demands of human dignity are written in every heart. ... Many of you will help achieve this better world. At a young age, you've taken up a great calling. You'll serve

> your country and our values. You'll protect your fellow citizens. And, by your effort and example, you will advance the cause of freedom around the world. And so I'm here to thank you for your commitment and congratulate you on the high honor you have received.
>
> (Bush, 17/4/2002)

Although each of these statements is important for their resonance with those we have considered above, Bush's second set of remarks here are particularly interesting for two further reasons. In the first instance, this statement offers a succinct combination of each of the detemporalising elements that we have identified as contributing to this second construction of timeless. It does so, by presenting the War on Terror as indicative of a perpetual fight for freedom, and then explicitly positioning this freedom (collocated here with dignity, respect, and tolerance) as a dehistoricised (and despatialised) moral absolute. Written in this way, there can be no question of this new conflict representing a purely American endeavour. The Germans, the Japanese, and, indeed, even the Russians had all come, finally, to realise the need to battle for human liberty throughout the world.

Second, and more importantly, this writing is also interesting for Bush's attempt to combine this specific writing of timelessness with two explicitly linear understandings of temporality. This introduction of linearity occurs in the first instance by drawing on the writings of progress and political certainty that we considered in Chapter 4. Not only are we convinced of this conflict's normative legitimacy here, we can also look to historical precedents for confirmation of its eventual outcome. And, in the second instance, this statement also approaches the existence and spread of freedom as an instrumentalist mechanism for the spreading of (the equally desirable normative values of) peace and security. Approached thus, as with the writing considered above, this fight to advance the cause of freedom was not merely justified on deontological ethical terms, where:

> There is a line in our time, and in every time, between those who believe all men are created equal, and those who believe that some men and women and children are expendable in the pursuit of power. There is a line in our time, and in every time, between the defenders of human liberty and those who seek to master the minds and souls of others.
>
> (Bush, 11/9/2002b)

It was justified, also, on the basis of its (knowable) ethical consequences (see also, for example, Bush, 17/10/2001, 25/3/2002, 17/7/2002, 8/3/2005; Powell, 5/10/2001; Rice, 25/8/2003). And, as Bush's subsequent remarks concerning Iraq further suggested, this linkage of liberty and peace provided considerable normative authority in some of the most controversial and contestable moments of this unfolding war: 'We have no ambition in Iraq except the liberation of its people. We ask no reward except a durable peace. And we will accept no outcome short of complete and final success' (Bush, 26/3/2003).

Although a particularly persistent and visible advocate of this second, cyclical, construction, Bush was far from alone in positioning the War on Terror thus (see also, for example, Bush, 11/11/2001, 21/11/2001, 7/12/2001, 4/5/2002). Vice President Cheney (2/10/2002), for example, similarly located the current campaign within a perpetual and enduring battle for freedom in his October 2002 remarks at the Washington Hilton:

> America is again called by history to use our overwhelming power in defense of our freedom. We've accepted that duty, certain of the justice of our cause, and confident of the victory to come. For my part, I am grateful for the opportunity to work with a president who is making us all proud, upholding the cause of freedom and serving the greatest nation on earth.

A theme also drawn out by Ashcroft (16/9/2002) in his earlier reflections on the consequences of 9/11 in Belgium:

> This week, as the world marks the anniversary of history's greatest assault on the American homeland, the United States and Europe remain linked by what bound us 18 years ago, and forty years before that. September 11 was a bitter reminder that our shared values – values of equality, justice, and freedom – are still under assault from some corners of the world. And like the freedom-loving soldiers who stormed the beaches of Normandy 58 years ago, we too have a duty to preserve and defend these values – at all costs.

Ashcroft's (1/10/2002) subsequent remarks to the USA Attorneys Conference in New York noted the difficulties of battling for liberty at this particular time:

> Despite the urgency of our responsibility, and the nobility of our labor, the historical defense of liberty has not always been easy, and

> it hasn't always been without controversy. In the midst of the Civil War, Abraham Lincoln worried aloud if there were not in all republics an 'inherent and fatal weakness', wondering whether democratic government must, of necessity, either be, quote, 'too strong for the liberties of its own people, or too weak to maintain its own existence?' History has answered this question. ... Today, America confronts a war against liberty like no other that we have even seen – it is a chaotic conflict that endangers not just our soldiers abroad but imperils our citizens at home; an enemy that lives among us, turning our freedoms into the means of freedom's destruction. In response to this new war, America has once again arisen to the defense of liberty.

For Condoleezza Rice, the War on Terror's location within this perpetual struggle for freedom even demonstrated the limitations of those arguments that had long forecast the obsolescence of major war in contemporary world politics (on this, compare Kaysen, 1990; Mueller, 1990; Mandelbaum, 1998):

> If you go back just a little while, there was a lot of speculation that the future of the American armed forces actually lay largely in so-called operations other than war, in policing civil and ethnic conflicts and in humanitarian missions. Well, I don't hear anybody saying that any more. The fact is that there are real threats out there. Freedom still faces ruthless enemies. And America and its allies have to deter and, if necessary, defeat them.
>
> (Rice, 1/2/2002)

A theme to which she returned in her subsequent press briefing on the President's forthcoming trip to Europe and Russia:

> This is a unique opportunity to remind all of us of how hard we fought together to defend freedom against tyranny in World War II. It reminds us that there are still threats to freedom, both from weapons of mass destruction and from terrorism. And it reminds us that what binds us as allies is our common beliefs and our common values and our willingness to defend them. And it reminds not us – not just us, but also our European allies, and it's gotten more expression since 9/11 in the way that our European allies have taken on that challenge.
>
> (Rice, 20/5/2002)

And more recently still, finally, Powell also compared the current conflict to the Cold War experience of the Eastern European nations in a manner akin to Bush's earlier 6 November 2001 remarks. The following statement is taken from his 27 July 2004 remarks at the Hungarian Chiefs of Mission Conference in Budapest:

> You've your own experience in this part of the world with struggles for freedom. From the dark days that shadowed this city in October of 1956, it took more than 30 years for your heroism to be vindicated, for Hungary to be free. It won't take that long for the hope in the hearts of the peoples of Afghanistan and Iraq to be vindicated, because the forces of freedom in the world are now stronger than ever. Hungary and its neighbors were once subjugated and tethered to the wrong side of history. Now you are free, and you are an integral and valuable part of the right side of history. We must not be faint-hearted in the face of current challenges. We must not waver or lose patience. We must stay the course for freedom in the course and in the face of danger.
>
> (Powell, 27/7/2004)

The politics of freedom versus fear

As we have seen, the Bush administration attempted persistently and repeatedly to position the War on Terror within a related conception of timelessness to that we considered above. As with the writing of good versus evil, this writing prioritised neither the emergent nor the evolutionary dynamics of this unfolding war. Rather, it conceptualised the War on Terror as part of a perpetual and ongoing dynamic of repetition, rebirth, and renewal. With time now (again) understood as a recurrent, strangely cyclical, process, this conflict was equivalent to, indeed repetition of, those wars for freedom that had been fought so successfully in earlier eras. Depicted in this way, World War II, the Cold War, the (not so new) War on Terror, and others besides, all here became part of one eternal dynamic: part of the same confrontation against those that would sow fear throughout the world. As Bush (21/11/2001) argued at a thanksgiving meal with his troops at Fort Campbell:

> Great causes are not easy causes. It was a long way from Bunker Hill to Yorktown. It was a long way for the 101st from Normandy to final victory over fascism in Europe. When wronged, our great nation has always been patient and determined and relentless. And that's the

> way we are today. We have defeated enemies of freedom before. And we will defeat them again.

In a very similar way to that considered above, this scripting of the War on Terror as the latest moment in a perpetual struggle for freedom afforded very little space for acknowledging earlier moments of American history that were eminently readable as a denial of, or challenge to, the liberties of certain individuals. Although a list of such moments affords scant recognition to their complexities, far greater justification would be required to demonstrate how the internment of 120,000 Japanese Americans during the World War II period (see Daniels, 2002) or the various McCarthyite policies of the 1940s and 1950s (see Schrecker, 2004), for example, could be legitimately incorporated into this cyclical historical narrative. Even more strikingly, this narrative again offered very little attention to the pernicious and pervasive practices of racial discrimination that long characterised American history: an exclusion explicitly acknowledged, indeed, by Condoleezza Rice on several occasions (4/12/2002, 31/10/2003). As she noted in her August 2003 remarks to Veterans of Foreign Wars:

> It is also true that democracy is not easy. Its institutions are not the natural embodiment of human nature. And our own history should remind us that the union of democratic principle and practice is always a work in progress. When the Founding Fathers said 'We the People', they did not mean me. My ancestors were considered three-fifths of a person. Knowing the difficulties of America's own history, we should always be humble in singing freedom's praises. But America's voice should never waver in speaking out on the side of people seeking freedom.
>
> (Rice, 25/8/2003)

More interesting, for our present purposes, was the way in which a second exclusion central to this writing of timelessness also interacted with the writing of moderated normality considered in the previous chapter. As we saw there, opposition to the War on Terror and its components were foreclosed not only by reference to the unprecedented, exceptional, security climate confronting America(ns) but also by appeal to the administration's efforts at protecting openness, liberty, and the (ostensibly bounded, ostensibly coherent) American way of life. With those same (ambiguous, vague, yet politically powerful) values now detached from any sense of time and space in this construction, we get a far clearer sense of the value of this scripting of timelessness for decontesting the War on Terror's very necessity and existence. For, as Bush (17/12/2005) argued in the approach to the expiry of the USA PATRIOT Act's 'sunset provisions', this act needed to be viewed not (only) as performing a narrow security function, but (also) as crucial to the perpetual, cyclical, struggle for freedom within which his administration was engaged:

> America's law enforcement personnel have used this critical law to prosecute terrorist operatives and supporters, and to break up terrorist cells in New York, Oregon, Virginia, California, Texas and Ohio. The Patriot Act has accomplished exactly what it was designed to do: it has protected American liberty and saved American lives.

Seen through this lens, indeed, even the abuse of prisoners incarcerated at the Abu Ghraib prison in Iraq could be swept aside, as but a temporary detour within this never-ending battle for liberty and dignity:

> Our nation has been through three difficult years in Iraq. And the way forward will bring more days of challenge and loss. The progress we've made has been hard-fought, and it's been incremental. There have been setbacks and missteps – like Abu Ghraib – that were felt immediately and have been difficult to overcome. Yet we have now reached a turning point in the struggle between freedom and terror.
> (Bush, 22/5/2006)

This writing of timelessness, in short, worked as a powerful attempt to exclude any form of critical resistance to the War on Terror, subsuming 'setbacks and misteps' that would occur – had occurred – along the way under a far more significant, a far more embedded, permanent antagonistic struggle.

To conclude our discussion now, it will be useful briefly to reflect on the functions performed by this second (contingent, partial, and violent) construction of timelessness. In the first instance, this writing, clearly, contributed to the inscription of significance into the unfolding War on Terror that we have traced throughout our study. In contrast to the discontinuous accounts of Chapter 3, and in common with the Gnostic writing explored directly above, this particular production of the War on Terror worked to position this conflict as significant *because of* its location within this recurrent struggle against the enemies of freedom. By depicting the rationale and conduct of this war as part of a recurrent dynamic far transcending the fleeting, strategic confrontation of its specificity, the Bush administration once again rendered the victorious undertaking of the War on Terror a historical imperative. As Bush's 20 September 2001 address to Congress indicated, the stakes of defeat went considerably further than the defence of mere life alone:

> These terrorists kill not merely to end lives, but to disrupt and end a way of life. With every atrocity, they hope that America grows fearful, retreating from the world and forsaking our friends. They stand against us, because we stand in their way.
>
> (Bush, 20/9/2001)

Produced in this way, opposition to this otherwise new war not only posed temporary (if significant) implications for the security of individuals and communities throughout America and beyond. It also represented a denial of the very freedom, the very way of life, which the United States (aided by others) had perpetually defended.

Second, this writing also worked to further inscribe the sense of normative certainty into the War on Terror that we encountered in the above writing of good versus evil. Once again, this production not only rendered this unfolding conflict significant, it also rendered it legitimate and just. By consistently defending the battle for freedom on both deontological grounds – as 'the universal longing of every soul' (Rice, 20/6/2005) and 'the birthright of every man, woman and child on earth' (Rumsfeld, 11/10/2001) – and, indeed, consequentialist grounds – 'a free Iraq and a free Afghanistan is a big step toward making the world more peaceful' (Bush, 10/8/2004) – the Bush administration rendered not fighting the War on Terror simply morally unthinkable. Failing to confront this latest challenge to freedom, in this sense, not only implied the potential – if important – loss of security and liberty considered above, it also implied a(nother) absolute moral transgression.

If this cyclical writing functioned to inscribe both significance and normative legitimacy into the unfolding War on Terror, its third implication was to add further structural coherence to the war itself. As with the good versus evil narrative, this writing again facilitated the Bush administration's incorporation of the various strategies and moments of this war within one single, overarching, rubric. By integrating otherwise disassociable developments such as military operations in Afghanistan (Bush, 1/5/2002; White House, 29/4/2003) and Iraq (Bush, 3/4/2003, 12/4/2003), the United States' broader foreign policy within the Middle East (Bush, 6/11/2003; White House, 6/11/2003), and the restructuring of domestic security policy (White House, 17/3/2003) into this singular discursive grammar constructed around the perpetual (and necessary) fight for freedom, the War on Terror once more took on the appearance of a consistent, if multifaceted, battle. If this conflict – as we noted in our introductory chapter – has often seemed complex and unwieldy to its observers and critics, it would have seemed far further complex and unwieldy in the absence of this latest of our temporal constructions.

And, finally, this second construction of timelessness also undoubtedly contributed to the processes of collective identity formation with which we have also been concerned throughout our discussion. By presenting the War on Terror not (only) as necessary for the defence of American national security, but, rather (or also), as part of an eternal ethical struggle, this writing again served to broaden this war's parameters and participants beyond the United States' borders. As earlier enemies had been transformed into contemporary allies by persuasion of the virtue of freedom, the world's (perpetual and recently converted) freedom-loving peoples and nations had a duty to convert contemporary enemies. Recognising and obeying freedom's call, in short, meant recognising this shared, universal, identity through confronting the shared threat posed by terrorists and their equivalents. A call to all nations in this war to save the world had been raised.

In sum, the Bush administration's writing of the War on Terror as part of a perpetual conflict between freedom and fear mobilised a similar conception of timelessness to the above construction of good versus evil. As with all of those we have considered thus far, it is crucial that we read it again as an inherently political construction: emphatically not as a neutral or inevitable reflection on this war. This is so, not only for its exclusion of other readings of American pasts, presents, and futures but also because of the further contribution it offered to the necessity, legitimacy, and identity politics of this war.

Conclusion

We embarked on our discussion in this chapter with a reminder that we have already encountered two radically distinct conceptions of temporality coursing through the War on Terror's construction. Claims to discontinuity and linearity, as we have seen, were both similarly central to this war's constitution and development: rendering it as a necessary, inevitable, and legitimate response to the events of 11 September 2001. What we have now, however, is yet another account of the movement and shape of time within this same context: an account that differs once more quite markedly from either of our earlier concerns. For, the production of this conflict as but the latest moment in a perpetual battle between good and evil, freedom and fear relied neither on assertions of rupture nor continuity. Instead, it was structured around a series of implicit and explicit claims to repetition and return: situating the War on Terror as a reincarnated, revisited, version of earlier struggles with fascism, communism, and any number of other historical undesirables.

By tracing the emergence and implications of these two writings of timelessness we have now a still fuller account of the War on Terror's construction around specific representations of time and history. In the first instance, we have seen that writings of timelessness performed a number of key functions for this pervasive and far-reaching conflict, not least in their

- positioning the contemporary threat of terrorism and its equivalents within a cycle of recurrent evil and hatred;
- legitimising the institution of the War on Terror by appeal to specific historical analogies;
- justifying the institution of potentially controversial policies and measures by depicting the contemporary terrorist enemy as an absolute moral other;
- opening discursive space for expanding the War on Terror's enactment and significance by both spatially and temporally unbounding this conflict;
- and introducing structural coherence into this war by linking its distinct moments as equivalent contributions to a perpetual historical struggle.

Second, we have also seen that these two writings of timelessness worked, once again, to exclude alternative – equally plausible – interpretations of the War on Terror, its actors, and import. And, by reflecting, finally, on this additional moment of discursive heterogeneity within this political discourse we are here now still further removed from any sense that it represented an objective or coherent response to the 9/11 attacks. If claims to time as a punctuated *and* directional dynamic worked to both legitimise and delegitimise particular violences in this conflict's unfolding, if claims to time as a punctuated *and* directional dynamic both called forth and banished specific identities within its development, then claims to time as a recurrent, timeless, and perpetual horizon did precisely the same. In this sense, if conceptions of timelessness offered an equally productive discursive resource in the battle with terrorism to those we encountered in Chapters 3 and 4, they were, in addition, of similarly political status.

6
Time, Violence, Identity, Politics

I began this monograph by making two related, but potentially separable, claims. In the first instance, I observed that exploring the functions and significance of specific writings of temporality within the Bush administration's War on Terror would tell us something rather interesting about that particular conflict. To truly understand the successful positioning of that war as a coherent, necessary, indeed legitimate, response to the events of 9/11, I argued, fully necessitated engaging with its scripting around particular claims to a privileged knowledge of time. In the second, I suggested that a careful exploration of these dynamics would also likely tell us something equally important about the relationships between time, violence, identity, and the political more broadly. The insights we may gain from our analysis of the War on Terror, I indicated, would quite possibly resonate far beyond the specific context with which we have been here concerned. In this concluding chapter, I would like now to expand on this observation by pulling together a number of the arguments we have been exploring in our discussion. But, in order to do so, it will be necessary to begin by first revisiting our more immediate focus of study.

Times of terror revisited

As we have seen throughout each of the preceding chapters, the War on Terror offers a complex and interesting topic for any engaged critical inquiry of the sort I have attempted to offer here. Although my efforts at conceptualisation and analysis may not be convincing to all observers of that conflict, I hope that two themes running throughout our discussion will be persuasive to many. The first of these relates to the absolute saturation of this 'first conflict of the twenty-first century' with claims to temporality. In discussing this discourse as a totality and its more particular developments, in reflecting on its emergence and linkages to other political confrontations, in writing this war's individual and collective identities: both self and other, and in speaking with such certainty of the security climate now confronting America(ns) and others, the George W. Bush administration returned continually and repeatedly to the passage, dynamics, and movement of time. It was simply impossible for them, it seemed, to reflect on the War on Terror's status and import without discussing – or imagining – its pasts, presents, and futures.

The second theme concerns the remarkably heterogeneous character of these governmental representations of time. If assertions of temporality were never far from this unfolding conflict, these assertions were certainly not marked by any semblance of consistency. As we have seen, discussions of temporal discontinuity accompanied by statements of rupture, breakage, and qualitative change; notions of temporal linearity associated with historical evolution, quantitative transformation, and knowable futures; and claims to timelessness predicated on a vision of recurrent and repeated pasts, presents, and futures, all of these shapes were turned to by the Bush administration in constructing this (new, continuing, or reincarnated) War on Terror. Where we began by suggesting that time was important in the framing of this war, it would, perhaps, be more accurate to argue that there existed a plurality of times running simultaneously across it: all of which were equally important.

By tracing and reading these distinct assertions of temporality, this book's most immediate purpose was to supplement existing understandings of the War on Terror itself. With the Bush administration's tenure having so recently concluded, we may not yet have the 'distance of history' that would be necessary to fully account for their successes and failures. What we do have, however, is a timely opportunity for reflecting on the ease with which a much lampooned and derided republican party could commit its own state and others to a costly, controversial, and seemingly interminable confrontation both at home and abroad. If going to war today is indeed a more difficult undertaking than it was in the past (Jackson, 2005: 121–2), observers of the Bush administration may not always have had this impression.

Now, we know from a number of existing studies (Makdisi, 2002; Silberstein, 2002; Beck, 2003; Edwards, 2004; Lazar & Lazar, 2004; Noon, 2004; Jackson, 2005) that the Bush administration's use and manipulation of provocative, emotive, language was central to their garnering support for – or dissipating resistance to – the War on Terror's inevitability and legitimacy. There is, of course, nothing new here: violence is always, has always been, rendered (im)possible or (un)desirable discursively. What I hope to have shown in this book, however, is that to fully grasp both the power and persuasiveness of this language it is crucial that we reflect on the disparate representations of temporality within it. For, once we read the administration's quite considerable output carefully and patiently, we rapidly encounter a far more liberal, far more extensive, series of temporal claims than one limited to governmental accounts of rupture and historical analogy: accounts of rupture and historical analogy that have – rightly – attracted so much criticism to date. And, taking these representations of time seriously is important, initially at least, for two reasons.

Most immediately, exploring the range of claims to time made by Bush and his administration serves to open up one further dimension of this conflict for critical analysis. If we are, really, interested in understanding and critiquing this war's existence and quite devastating impacts, we are behoved, I believe, to garner as full a conception of it as is possible. Although we need resist the temptation to speak the truth on this (new, continuing, recurrent) War on Terror, bringing forth hitherto hidden, hitherto camouflaged, aspects of this (incomplete, political) totality is an imperative for truly critical, truly responsible, scholarship. For, as Ricoeur (1999: 16) argues in a rather different context:

> The fundamental objective of the good historian is to enlarge the sphere of archives; that is, the conscientious historian must open up the archive by retrieving traces which the dominant ideological forces attempted to suppress. In admitting what was originally excluded from the archive the historian initiates a critique of power. ... The historian opposes the manipulation of narratives by telling the story differently and by providing a space for the confrontation between opposing testimonies.

Although we have not been directly concerned with the form of subjugated knowledges alluded to here by Ricoeur – and that elsewhere dominate Foucault's interests – what we have done is bring forth for scrutiny the centrality of temporality within this particular context. And this, I suspect, represents one productive way of telling the War on Terror story rather differently.

In the second instance, by focusing on the heterogeneity of these distinct claims to temporality, we also get a rather different understanding of the War on Terror's coherence or stability than that which was offered up by Bush's administration. That events and identities such as 9/11, America(ns), al-Qaeda, and the War on Terror as a totality proved capable of insertion into such different notions of time – simultaneously, and by the very same members of the administration – cannot but lead us to question the very objectivity of this extensive and dynamic discursive framework. By telling the War on Terror story differently, in this sense, not only do we shed light on elements of it that may have seemed concealed or secreted away to observers, participants, and analysts alike. We also, crucially, begin to understand that there is no – there can be no – War on Terror story in the singular. Its existence, inevitability, necessity, legitimacy, *and* coherence were all fashioned discursively – frequently through appeal to the temporal. Its existence, inevitability, necessity, legitimacy, and coherence, put otherwise, were all *made* – and emphatically not given. And, as we know from Foucault (1988: 154–5), a redescription of this sort aimed at rendering the seemingly inevitable contingent is far from a detached, neutral enterprise:

> A critique is not a matter of saying that things are not right as they are. It is a matter of pointing out on what kinds of assumptions, what kinds of familiar, unchallenged, unconsidered modes of thought the practices that we accept rest. ... Criticism is a matter of flushing out that thought and trying to change it: to show that things are not as self-evident as one believed, to see that what is accepted as self-evident will no longer be accepted as such. Practicing criticism is a matter of making facile gestures difficult.

Time and violence

Let us turn, now, to the broader relationships that exist between temporality, violence, identity, and the political that we pointed to at the start of our discussion. To begin, as we have seen throughout this book, temporality – or, better, claims to temporality – appears intimately wrapped up with that issue which has long been central to the disciplines of political science and International Relations alike: the exercise of violence. In the context of the War on Terror specifically, this linkage has emerged in at least three different ways.

In the first instance, claims to temporality functioned quite visibly, quite explicitly, to render certain violences both necessary and legitimate in this unfolding war. Whether justifying those violences through assertions of radical transformation, as in Paul Wolfowitz's (2/12/2002) vision of future catastrophe:

> I think it goes back to some, an understanding, of just how profoundly the events of September 11th changed America's understanding of the risk and the stakes ... we can't wait until 30,000 Americans or 300,000 or even possible 3000,000 die as a result of an attack by weapons of mass destruction to deal with the threat posed by countries that have the weapons and develop them and support and work with terrorists.

Or through the more distinctively linear efforts at forging pasts, presents, and futures into a coherent story of warnings finally heeded, lessons finally learned, as in Rumsfeld's (27/3/2006) more recent reflections on the years preceding 9/11: 'During those decades the West was ambivalent about how to counter extremist ideology and that type of aggression. As a result, terrorists became increasingly bolder [*sic*].' Specific representations of time were absolutely fundamental to the legitimisation of present/future violence and war by the Bush administration.

At the same time, interestingly, claims to temporality also offered a remarkably productive discursive resource for *delegitimising* certain violences: violences conducted, of course, by threatening others. Most obvious, perhaps, in the assertions to perpetual truths we encountered in our discussion of timelessness: 'It is always, and everywhere, wrong to target and kill the innocent. It is always, and everywhere, wrong to be cruel and hateful, to enslave and oppress' (Bush, 7/2/2002); time could be usefully employed – indeed, was usefully employed – to render the killing of some, by some, morally impermissible.

And, finally, perhaps most important of all, we have also seen that particular readings of pasts, presents, and futures could further be useful in *erasing or excluding* certain violences from stories that were under construction. On the one hand, we encountered the ways in which abuses and mistreatments could be at once invoked and passed over as mere delays in the unfolding of more pertinent temporal dynamics: 'There have been setbacks and missteps – like Abu Ghraib – that were felt immediately and have been difficult to overcome. Yet we have now reached a turning point in the struggle between freedom and terror'

(Bush, 22/5/2006). On the other, we have also seen the difference that alternative readings of pasts, presents, and futures – readings acknowledging, for example, structural violences occurring throughout the world – would have made to any of the claims of uniqueness, continuity, and timelessness that we have been tracing. In either case, representations of temporality worked not only to call forth particular violences for the acceptance or condemnation of audiences and observers. They also, crucially, worked to obscure alternative violences that may otherwise have been quite legitimately invoked or considered.

There is, of course, a need to caution against the desire towards extravagant claims on the basis of this one study alone. That said, the productivity and resonance of claims to time in facilitating, foreclosing, and concealing particular violences appear far from unique to this War on Terror. In the contemporary period, for example, we have witnessed a range of practices designed to revisit and reread the past differently as a mechanism of retribution or compensation for historical injustices and violences: most visibly in demands for executive public apologies and in the activities of the recent Truth and Reconciliation Commissions in South Africa and beyond. And, the couching of military violence within assertions of threat, risk, and danger that has long interested scholars of global politics also cannot but rely on an appeal to specific future scenarios that may or may not depart from the past and present. If constructions or performances of temporality are not necessarily, inevitably, present within the unfolding or avoidance of particular violences, then, time may always be – can always be – employed to such ends.

Time and identity

In Chapter 2 we encountered a range of literatures that have together drawn our attention to time's existence not as the objective, extra-discursive, entity we may typically imagine it to be, but, rather, as a narrative or social production that poses significant, tangible, implications for the lives of individuals, groups, and communities. Where these literatures do not constitute a coherent or unitary body of work, they do frequently share an attempt to explore the connections between narratives or performances of time and the emergence of specific identity practices at particular historical junctures.

In one sense, each of the preceding chapters offers an attempt to contribute to this common concern. As we have seen, America, Americans, civilisation, the world, and, indeed, many of the other collective identities that were drawn on by the Bush administration in their writing of the War on Terror were produced precisely through an explicit appeal to temporality. Most obviously, of course, this took place within those invocations of shared pasts we have noted: be they histories of confronting terrorism, histories of fighting bravely for freedom and goodness against fear and evil, or even the rather less glorious history of self-interest and hedonism we saw in Chapter 3's writing of self-transformation. In each case, who 'we' were – what 'we' were – was scripted around an imaginary account of either unity or sameness at an earlier temporal juncture. On top of this, however, we have also encountered a number of ways in which invocations of common presents and futures performed the very same interpellative functions: whether in reference to present/futures of threat, self-sacrifice, family values, or global peace. And, as noted throughout our discussion, the administration's ability and willingness to imagine *distinct* collective self-identities in the past, present, and/or future was frequently integral to their efforts at legitimising the expansion of this conflict beyond the American borders.

If we have encountered a connection between time and self-identity in the writing of the War on Terror, we have also come across something quite similar in the constructions of threat, danger, and otherness that were equally central to this conflict's borders and rationale. In representations of terrorists and their equivalents as somehow more deadly, more threatening, than earlier, deterrable, rational adversaries: 'you should not wait to be surprised by evil people who may wish you real harm with weapons of mass destruction that would make September 11 look small in comparison' (Rice, 8/9/2002). Or, attempts to position these threats at the cusp of an escalating history of unconventional violence: 'Al Qaeda had previously demonstrated the ability to strike again; to carry out multiple, complex and simultaneous attacks' (Ashcroft, 9/9/2003). Or even, finally, in claims to these actors as repetition of earlier, equally bloodthirsty, amoral foes: 'Like the fascists and totalitarians before them, these terrorists – al Qaeda, the Taliban regime that supports them, and other terror groups across our world – try to impose their radical views through threats and violence' (Bush, 6/11/2001). In each case, again, who 'they' were – what 'they' were – was persistently, if not exclusively, determined by 'their' location within particular temporal trends.

To fully understand the identity politics of both self and other in the War on Terror, then, it seems clear that we cannot but take seriously its writing around specific conceptions of time and history. If claims to a privileged knowledge of the temporal worked to legitimise, delegitimise, and conceal particular violences as we have seen, claims to a privileged knowledge of the temporal also performed a crucial constitutive role in framing both them and us in this context.

Time and the political

The final, and most urgent, theme we need to explore here builds directly on each of the above, and concerns the relationship between time and the political. As we know from existing studies, time is bound up with the political in at least two different respects. In the first instance, most visibly, Gross (1985) has pointed to the ways in which modern states systematically appropriated the mechanisms and symbols that are central to measuring, marking, and controlling temporality from the nineteenth century onwards particularly. By replacing, or attempting to replace, those religious and traditional events that had previously loomed large in individual orientations towards time and history with national holidays and the like, a distinctively secular, distinctively linear, conception of temporality began to take precedence in social consciousness over other, earlier, forms of periodisation. Facilitated through the nationalisation of mass education, this new – rationalised – approach to thinking and experiencing time posed profound implications for the identities of subjects and citizens. For, as Gross (1985: 68) notes: 'The net effect of this, at least from the state's perspective, would undoubtedly be a more manageable polity, with more social order, more cohesion, more homogeneity.'

From a slightly different perspective, Nowotny (1994: 102–7) has suggested that time possesses an inherently political character in part because of its scarcity. Once we realise that time – my time, your time, our time – is a finite phenomenon, we can see that it is subject to the familiar problems of distribution, exchange, and ultimately conflict that characterise all political entities and values. Time, at least in the contemporary period, has become a resource that is at once susceptible to management, productive (or non-productive) employment, and trade. Who determines how our time is used and to what purposes, from such a perspective, becomes, necessarily, a question of politics. If Harold Lasswell's famous framing of the political as pertaining to who gets what, when, and how continues to be pertinent, it is, perhaps, nowhere more so than with respect to time.

The approach taken in this book has been to sketch a slightly different way again of conceptualising the relationship between time and the political. In the first instance, as we have seen, assertions of a privileged knowledge of time that were made by key representatives of the Bush administration in their writing of the War on Terror performed a radically constitutive role in bringing this conflict into being. These assertions not only worked to legitimise or justify this (new, continuing, or reincarnated) war's very emergence and presence, they also shaped its parameters and conduct. Military interventions, treatments of terrorist suspects, homeland security measures, all of these and many others besides were *made possible* by the manipulation, imagination, and scripting of temporality. If the political is associated with those moments of institution that produce and reproduce our social worlds (Laclau, 1996: 47; Edkins, 1999: 2, 2003: 13–14), then, time's incredible capacity to perform this instituting function renders it, I suggest, inherently political.

The second sense in which claims to time have been seen as political throughout our discussion relates to the exclusionary role that they performed in concealing, foreclosing, or militating against alternative aspects of the War on Terror itself. Representations of temporal discontinuity, linearity, and timeless, as we have noted, not only coerced against alternative, equally plausible, understandings of the movement or shape of time in the object being narrated: be that America(ns), terrorism, the War on Terror and so forth. They also, simultaneously, coerced against a potentially infinite array of equally plausible responses to those objects that were being positioned within these temporal shapes. To take one example, as we saw in Chapter 3, the Bush administration worked particularly hard to position 9/11 as an exceptional, unprecedented, moment of violence. This positioning not only excluded those efforts to understand the attacks differently – as part of a linear or timeless dynamic that we encountered in our subsequent chapters – it also, more pertinently, worked to foreclose alternative – unexceptional – strategies for responding to the attacks themselves: at least in their most immediate aftermath. In this sense, where White (1990: 62–3) and Mouffe (2005: 10–14), for example, have tied the political to the practices of exclusion, othering, and the production of difference that run through academic disciplines and social worlds respectively, the imagination or representation of time offers a further particularly vivid example of this same dynamic. It is simply not possible to speak the truth on time from any position of certainty or assurance without, simultaneously, denying other temporal truths. And, for these reasons, this book stands as a call to afford greater attention to the ways in which time is shaped or narrated. Not only in these most recent *Times of Terror*, but, indeed, also beyond.

Bibliography

Agger, B. (1991) 'Critical Theory, Poststructuralism, Postmodernism: Their Sociological Relevance', in *Annual Review of Sociology* 17(1): 105–31.

Ahmad, M. (2002) 'Homeland Insecurities: Racial Violence the Day after September 11', in *Social Text 72* 20(3): 101–15.

Alexander, G. (2001) 'Institutions, Path Dependence, and Democratic Consolidation', in *Journal of Theoretical Politics* 13(3): 249–69.

Alheit, P. (1994) 'Everyday Time and Life Time: On the Problems of Healing Contradictory Experiences of Time', in *Time & Society* 3(3): 305–19.

An-Na'im, A. (2002) 'Upholding International Legality against Islamic and American *Jihad*', in K. Booth, & T. Dunne (eds) *Worlds in Collision: Terror and the Future of Global Order*. Basingstoke: Palgrave Macmillan, pp. 162–71.

Aretxaga, B. (2002) 'Terror as Thrill. First Thoughts on the 'War on Terrorism'', in *Anthropological Quarterly* 75(1): 139–50.

Ashley, R. (1989) 'Living on Border Lines: Man, Poststructuralism, and War', in J. Der Derian & M. Shapiro (eds) *International/Intertextual Relations: Postmodern Readings of World Politics*. New York: Lexington Books, pp. 259–320.

Ashley, R. & Walker, R. (1990) 'Conclusion: Reading Dissidence/Writing the Discipline: Crisis and the Question of Sovereignty in International Studies', in *International Studies Quarterly* 34(3): 367–416.

Baker, N. (2003) 'National Security versus Civil Liberties', in *Presidential Studies Quarterly* 33(3): 547–67.

Bash, H. (2000) 'A Sense of Time: Temporality and Historicity in Sociological Inquiry', in *Time & Society* 9(2/3): 187–204.

Bearman, P., Faris, R. & Moody, J. (1999) 'Blocking the Future: New Solutions for Old Problems in Historical Social Science', in *Social Science History* 23(4): 501–33.

Beck, U. (2003) 'The Silence of Words: On Terror and War', in *Security Dialogue* 34(3): 255–67.

Bellamy, A. (2006). 'No pain, No Gain? Torture and Ethics in the War on Terror', in *International Affairs* 82(1): 121–48.

Bellamy, A. (2008) 'Pre-empting Terror', in A. Bellamy, R. Bleiker, S. Davies & R. Devetak (eds) *Security and the War on Terror*. London: Routledge, pp. 106–21.

Bellamy, A., Bleiker, R., Davies, S. & Devetak, R. (eds) (2008) *Security and the War on Terror*. London: Routledge.

Bergmann, W. (1992) 'The Problem of Time in Sociology: An Overview of the Literature on the State of Theory and Research on the "Sociology of Time", 1900–82', in *Time & Society* 1(1): 81–134.

Booth, K. (2008) 'The Human Faces of Terror: Reflections in a Cracked Looking-Glass', in *Critical Studies on Terrorism* 1(1): 65–79.

Borradori, G. (2003) 'Deconstructing Terrorism – Derrida', in G. Borradori (ed.) *Philosophy in a Time of Terror: Dialogues with Jürgen Habermas and Jacques Derrida*. London: The University of Chicago Press, pp. 137–72.

Bousequet, A. (2006) 'Time Zero: Hiroshima, September 11 and Apocalyptic Revelations in Historical Consciousness', in *Millennium: Journal of International Studies* 34(3): 739–64.
Brockmeier, J. (1997) 'Autobiography, Narrative, and the Freudian Concept of Life History', in *Philosophy, Psychiatry, & Psychology* 4(3): 175–99.
Bruner, J. (1991) 'The Narrative Construction of Reality', in *Critical Inquiry* 18(1): 1–21.
Burnett, J. & Whyte, D. (2005) 'Embedded Expertise and the New Terrorism', in *Journal for Crime, Conflict and the Media* 1(4): 1–18.
Butler, J. (2006) *Precarious Life: The Powers of Mourning and Violence*. London: Verso.
Campbell, D. (1998a) *National Deconstruction: Violence, Identity, and Justice in Bosnia*. London: University of Minnesota Press.
Campbell, D. (1998b) *Writing Security: United States Foreign Policy and the Politics of Identity* (2nd Edn). Manchester: Manchester University Press.
Campbell, D. (2002) 'Time is Broken: The Return of the Past in the Response to September 11', in *Theory and Event*: 5(4).
Caputo, J. (1997) *Deconstruction in a Nutshell: A Conversation with Jacques Derrida*. New York: Fordham University Press.
Charteris-Black, J. (2005) *Politicians and Rhetoric: The Persuasive Power of Metaphor.* Basingstoke: Palgrave Macmillan.
Chatman, S. (1981) 'Reply to Barbara Herrnstein Smith', in *Critical Inquiry* 7(4): 802–9.
Chomsky, N. (2003) *Hegemony or Survival: America's Quest for Global Dominance*. London: Hamish Hamilton.
Critchley, S. & Marchart, O. (2004) 'Introduction', in S. Critchley & O. Marchart (eds) *Laclau: A Critical Reader*. London: Routledge, pp.1–14.
Clough, P. (1991) 'Poststructuralism and Postmodernism: The Desire for Criticism', in *Theory and Society* 21(4): 543–52.
Cole, D. (2002) 'Enemy Aliens', in *Stanford Law Review* 54(5): 953–1004.
Cole, D. (2003) *Enemy Aliens*. New York, NY: New Press.
Collins, J. (2002) 'Terrorism', in J. Collins & R. Glover, (eds) *Collateral Language: A User's Guide to America's New War*. London: New York University Press, pp. 155–74.
Collins, J. & Glover, R. (eds) *Collateral Language: A User's Guide to America's New War*. London: New York University Press.
Cox, M. (2002) 'American Power Before and After September 11th: Dizzy with Success?', in *International Affairs* 78(2): 261–76.
Cox, R. (1996) 'Social Forces, States, and World Orders: Beyond International Relations Theory (1981)', in R. Cox, (ed.) *Approaches to World Order*. Cambridge: Cambridge University Press, pp.85–123.
Crawford, N. (2003) 'Just War Theory and the U.S. Counterterror War', in *Perspective on Politics* 1(1): 5–25.
Croft, S. (2006) *Culture, Crisis and America's War on Terror*. Cambridge: Cambridge University Press.
Cronon, W. (1992) 'A Place for Stories: Nature, History and Narrative', in *The Journal of American History* 78(4): 1347–76.
Cunliffe, A., Luhman, J. & Boje, D. (2004) 'Narrative Temporality: Implications for Organizational Research', in *Organization Studies* 25(2): 261–86.

Daniels, R. (2002) 'Incarceration of the Japanese-Americans: A Sixty-Year Perspective', in *The History Teacher* 35(3): 297–310.

Davies, K. (2001) 'Responsibility and Daily Life: Reflections over timespace', in J. May & N. Thrift (eds) *Timespace: Geographies of Temporality*. London: Routledge, pp.133–48.

Der Derian, J. (2002) '*In Terrorem*: Before and After 9/11', in K. Booth & T. Dunne (eds) *Worlds in Collision: Terror and the Future of Global Order*. Basingstoke: Palgrave Macmillan, pp. 101–17.

Der Derian, J. (2003) 'Decoding the *National Security Strategy of the United States of America*', in *Boundary 2* 30(3): 19–27.

Derrida, J. (1997) 'The Villanova Roundtable: A Conversation with Jacques Derrida', in J. Caputo (ed.) *Deconstruction in a Nutshell: A Conversation with Jacques Derrida*. New York: Fordham University Press, pp. 3–28.

Derrida, J. (2003) 'Autoimmunity: Real and Symbolic Suicides', in G. Borradori (ed.) *Philosophy in a Time of Terror: Dialogues with Jurgen Habermas and Jacques Derrida*. London: The University of Chicago Press, pp. 85–136.

Desai, G. (1993) 'The Invention of Invention', in *Cultural Critique* 24: 119–42.

Devetak, R. (2005) 'The Gothic Scene of International Relations: Ghosts, Monsters, Terror and the Sublime after September 11', in *Review of International Studies* 31(4): 621–43.

Dittmer, J. (2005) 'Captain America's Empire: Reflections on Identity, Popular Culture and Post-9/11 Geopolitics', in *Annals of the Association of American Geographers* 95(3): 626–43.

Duyvesteyn, I. (2004) 'How New is the New Terrorism?', in *Studies in Conflict and Terrorism* 27(5): 439–54.

Edkins, J. (1999) *Poststructuralism and International Relations: Bringing the Political Back In*. London: Lynne Rienner.

Edkins, J. (2003) *Trauma and the Memory of Politics*. Cambridge: Cambridge University Press.

Edwards, J. (2004) 'After the Fall', in *Discourse and Society* 15(2–3): 155–84.

Elchardus, M. (1988) 'The Rediscovery of Chronos: The New Role of Time in Sociological Theory', in *International Sociology* 3(1): 35–59.

Falk, R. (2002) 'Testing Patriotism and Citizenship in the Global Terror War', in K. Booth & T. Dunne (eds) *Worlds in Collision: Terror and the Future of Global Order*. Basingstoke: Palgrave Macmillan, pp. 325–35.

Feldman, A. (1991) *Formations of Violence: The Narrative of the Body and Political Terror in Northern Ireland*. London: The University of Chicago Press.

Fortier, A. (1999) 'Re-membering Places and the Performance of Belonging(s), in *Theory, Culture & Society* 16(2): 41–64.

Foucault, M. (1981) 'The Order of Discourse', in R. Young (ed.) *Untying the Text: A Poststructuralist Reader*. London: Routledge, pp. 48–77.

Foucault, M. (1984) 'Polemics, Politics, and Problemizations: An Interview with Michel Foucault', in P. Rabinow (ed.) *The Foucault Reader: An Introduction to Foucault's Thought*. London: Penguin, pp. 381–90.

Foucault, M. (1988) 'Practicing Criticism', in L. Kritzman (ed.) *Michel Foucault: Politics, Philosophy, Culture: Interviews and Other Writings 1977–1984*. London: Routledge, pp. 152–6.

Foucault, M. (1990) 'On Power', in D. Kritzman (ed.) *Michel Foucault: Politics, Philosophy, Culture: Interviews and Other Writings 1977–1984*. London: Routledge, pp. 96–109.
Foucault, M. (2002) *The Order of Things: An Archaeology of the Human Sciences*. London: Routledge.
Freedman, L. (2002) 'The Third World War?', in *Survival* 43(4): 61–88.
Fukuyama, F. (1992) *The End of History and the Last Man*. London: Penguin.
Fukuyama, F. (1995) 'Reflections on the End of History, Five Years Later', in *History and Theory* 34(2): 27–43.
Galtung, J. (1969) 'Violence, Peace, and Peace Research', in *Journal of Peace Research* 6(3): 167–91.
Galtung, J. & Höivik, T. (1971) 'Structural and Direct Violence: A Note on Operationalization', in *Journal of Peace Research* 8(1): 73–6.
Ghandi, R. (2005) 'Closing Chapters of Enmity', in A. Ahmed, A. Ahmed & B. Forst (eds) *After Terror: Promoting Dialogue among Civilizations*. Cambridge: Polity Press, pp. 29–35.
Gilpin, R. (1988) 'The Theory of Hegemonic War', in *Journal of Interdisciplinary History* 18(4): 591–613.
Glennie, P. & Thrift, N. (1996) 'Reworking E. P. Thompson's "Time, Work Discipline and Industrial Capitalism"', in *Time & Society* 5(3): 275–99.
Glover, R. (2002) 'The War on ___', in J. Collins & R. Glover (eds) *Collateral Language: A User's Guide to America's New War*. London: New York University Press, pp. 207–22.
Gordon, A. (1999) 'Terrorism Dissertations and the Evolution of a Speciality: An Analysis of Meta-Information', in *Terrorism and Political Violence* 11(2): 141–50.
Gould, J. (2002) 'Playing with Fire: The Civil Liberties Implications of September 11th', in *Public Administration Review* 62(1): 74–9.
Graham, G. (1997) *The Shape of the Past: A Philosophical Approach to History*. Oxford: Oxford University Press.
Graham, P., Keenan, T. & Dowd, A. (2004) 'A Call to Arms at the End of History: A Discourse-Historical Analysis of George W. Bush's Declaration of War on Terror', in *Discourse & Society* 15(2–3): 199–221.
Gross, D. (1985) 'Temporality and the Modern State', in *Theory and Society* 14(1): 53–82.
Gunning, J. (2007) 'A Case for Critical Terrorism Studies?', in *Government and Opposition* 42(3): 363–93.
Hajer, M. (2005) 'Coalitions, Practices, and Meaning in Environmental Politics: From Acid Rain to BSE', in D. Howarth & J. Torfing (eds) *Discourse Theory in European Politics: Identity, Policy and Governance*. Basingstoke: Palgrave Macmillan, pp. 297–315.
Horgan, J. & Boyle, M. (2008) 'A Case against "Critical Terrorism Studies"', in *Critical Studies on Terrorism* 1(1): 51–64.
Howarth, D. (2000) *Discourse*. Buckingham: Open University Press.
Huddy, L., Feldman, S., Lahav, G. & Taber, C. (2003) 'Fear and Terrorism: Psychological Reactions to 9/11', in P. Norris, M. Kern & M. Just (eds) *Framing Terrorism: The News Media, the Government, and the Public*. London: Routledge, pp. 255–78.

IBC (2008) *Iraq Body Count*. Available at: http://www.iraqbodycount.org/ Accessed: 17/7/2008.
Isin, E. & Rygiel, K. (2007) 'Abject Spaces, Frontiers, Zones, Camps', in E. Dauphinee & C. Masters (eds) *The Logics of Biopower and the War on Terror: Living, Dying, Surviving*. Basingstoke: Palgrave Macmillan, pp. 181–203.
Ivie, R. (2005) *Democracy and America's War on Terror*. Tuscaloosa: The University of Alabama Press.
Jackson, R. (2005) *Writing the War on Terrorism: Language, Politics and Counter-Terrorism*. Manchester: Manchester University Press.
Jackson, R. (2007a) 'Responses', in *International Affairs* 83(1): 165–177.
Jackson, R. (2007b) 'Constructing enemies: 'Islamic Terrorism' in Political and Academic Discourse', in *Government and Opposition* 42(3): 394–426.
Jarvis, L. (2008) 'Times of Terror: Writing Temporality into the War on Terror', in *Critical Studies on Terrorism* 1(2): 245–62.
Jarvis, L. (2009) 'The Spaces and Faces of Critical Terrorism Studies', in *Security Dialogue* 40(1): 5–27.
Joas, H. (1999) 'The Modernity of War: Modernization Theory and the Problem of Violence', in *International Sociology* 14(4): 457–72.
John, P. (1999) 'Ideas and Interests; Agendas and Implementation: An Evolutionary Explanation of Policy Change in British Local Government Finance', in *British Journal of Politics and International Relations* 1(1): 39–62.
Johnson, N. (2001) 'From Time Immemorial: Narratives of Nationhood and the Making of National Space', in J. May & N. Thrift (eds) *Timespace: Geographies of Temporality*. London: Routledge, pp. 89–105.
Josselson, R. (2004) 'The Hermeneutics of Faith and the Hermeneutics of Suspicion', in *Narrative Inquiry* 14(1): 1–28.
Kahn, R. (1989) 'Women and Time in Childbirth and During Lactation', in F. Forman (ed.) *Taking Our Time: Feminist Perspectives on Temporality*. Oxford: Pergamon Press, pp. 20–36.
Kaysen, C. (1990) 'Is War Obsolete?: A Review Essay', in *International Security* 14(4): 42–64.
Kennedy, L. & Lucas, S. (2005) 'Enduring Freedom: Public Diplomacy and U.S. Foreign Policy', in *American Quarterly* 57(2): 309–33.
Kennedy, P. (1991) 'On the "Natural Size" of Great Powers', in *Proceedings of the American Philosophical Society* 135(4): 485–9.
Kennedy-Pipe, C. & Rengger, N. (2006) 'Apocalypse Now? Continuities or Disjunctions in World Politics after 9/11', in *International Affairs* 82(3): 539–52.
Kerr, P. (2002) 'Saved from Extinction: Evolutionary Theorising, Politics and the State', in *British Journal of Politics and International Relations* 4(2): 330–58.
Klein, N. (2002) 'Signs of the Times', in P. Scraton (ed.) *Beyond September 11: An Anthology of Dissent*. London: Pluto Press, pp. 143–9.
Kreiswirth, M. (2000) 'Merely Telling Stories? Narrative and Knowledge in the Human Sciences', in *Poetics Today* 21(1): 293–318.
Kristeva, J. (1981) 'Women's Time', in *Signs* 7(1): 13–35.
Kroes, P. (1984) 'Objective versus Minddependent Theories of Time Flow', in *Synthese* 61(3): 423–46.
Kuhn, T. (1996) *The Structure of Scientific Revolutions* (3rd Edn). London: The University of Chicago Press.

Kushner, H. (2002) 'Editors Foreword', in H. Kushner (ed.) *Essential Readings on Political Terrorism: Analyses of Problems and Prospects for the 21st Century*. Lincoln, NE: Gordian Knot Books, pp. xii–xv.

Lacina, B. & Gleditsch, N. (2005) 'Monitoring Trends in Global Combat: A New Dataset of Battlefield Deaths', in *European Journal of Population* 21(2–3): 145–66.

Laclau, E. (1989) 'Politics and the Limits of Modernity', in A. Ross (ed.) *Universal Abandon: The Politics of Postmodernism*. Edinburgh: Edinburgh University Press, pp. 63–82.

Laclau, E. (1990) *New Reflections on the Revolution of Our Time*. London: Verso.

Laclau, E. (1994) 'Introduction', in E. Laclau (ed.) *The Making of Political Identities*. London: Verso, pp.1–8.

Laclau, E. (1996) 'Deconstruction, Pragmatism, Hegemony', in C. Mouffe (ed.) *Deconstruction and Pragmatism*. London: Routledge, pp. 47–67.

Laclau, E. (2005) *On Populist Reason*. London: Verso.

Laclau, E. & Mouffe, C. (1987) 'Post-Marxism without Apologies', in *New Left Review*: I (166): 79–106.

Laclau, E. & Mouffe, C. (2001) *Hegemony and Socialist Strategy: Towards a Radical Democratic Politics* (2nd Edn). London: Verso.

Lakoff, G. (2001) 'Metaphors of Terror' Available at: http://www.press.uchicago.edu/News/911lakoff.html Accessed: 18.7.2005.

Laqueur, W. (1996) 'Postmodern Terrorism: New Rules for an Old Game', in *Foreign Affairs* 75(5): 24–36.

Laqueur, W. (1999) *The New Terrorism: Fanaticism and the Arms of Mass Destruction*. Oxford: Oxford University Press.

LaRossa, R. (1983) 'The Transition to Parenthood and the Social Reality of Time', in *Journal of Marriage and the Family* 45(3): 579–89.

Layne, C. (1993) 'The Unipolar Illusion: Why New Great Powers Will Rise', in *International Security* 17(4): 5–51.

Lazar, A. & Lazar, M. (2004) 'The Discourse of the New World Order: "Out-casting" the Double Face of Threat', in *Discourse and Society* 15(2–3): 223–42.

Leccardi, C. (1996) 'Rethinking Social Time: Feminist Perspectives', in *Time & Society* 5(2): 169–86.

Lee, R. & Barton, M. (2004) 'September 11, 2001 and the Rituals of Religious Pluralism', in R. Denton (ed.) *Language, Symbols and the Media: Communication in the Aftermath of the World Trade Center Attack*. London: Transaction Publishers, pp. 17–48.

Lenzo, K. (1995) 'Validity and Self-Reflexivity Meet Poststructuralism: Scientific Ethos and the Transgressive Self', in *Educational Researcher* 24(4): 17–23.

Lewis, J. (2005) *Language Wars: The Role of Media and Culture in Global Terror and Political Violence*. London: Pluto Press.

Lewis, J. & Weigert, A. (1981) 'The Structures and Meanings of Social Time', in *Social Forces* 60(2): 432–62.

Lukes, S. (2006). 'Liberal Democratic Torture', in *British Journal of Political Science* 36(1): 1–16.

Lyon, D. (2003) *Surveillance after September 11*. London: Polity.

Lyotard, J. (1984) *The Postmodern Condition: A Report on Knowledge*. Manchester: Manchester University Press.

Mahoney, J. (2000) 'Path Dependence in Historical Sociology', in *Theory and Society* 29(4): 507–48.
Makdisi, S. (2002) 'Spectres of "Terrorism"', in *Interventions*: 4(2): 265–78.
Mandelbaum, M. (1998) 'Is Major War Obsolete?', in *Survival* 40(4): 20–38.
McFarland, A. (1991) 'Interest Groups and Political Time: Cycles in America', in *British Journal of Political Science*: 21(3): 257–84.
McInnes, C. (2003) 'A Different Kind of War? September 11 and the United States' Afghan War', in *Review of International Studies* 29(2): 165–84.
Meisels, T. (2005) 'How Terrorism Upsets Liberty', in *Political Studies*: 53(1): 162–81.
Merari, A. (1991) 'Academic Research and Government Policy on Terrorism', in *Terrorism & Political Violence* 3(1): 88–102.
Migration Policy Institute (2003) *America's Challenge: Domestic Security, Civil Liberties, and National Unity after September 11*. Available at: http://www.migrationpolicy.org/pubs/Americas_Challenges.pdf Accessed: 3.1.2006.
Miller, W. (2000) 'Durkheimian Time', in *Time & Society* 9(1): 5–20.
Mills, S. (2004) *Discourse* (2nd Edn). London: Routledge.
Modelski, G. (1978) 'The Long Cycle of Global Politics and the Nation-State', in *Comparative Studies in Society and History* 20(2): 214–35.
Morris, R. (2004) 'Images of Untranslatibility in the US War on Terror', in *Interventions* 6(3): 401–23.
Mouffe, C. (2005) *On the Political*. London: Routledge.
Mueller, J. (1990) 'The Obsolescence of Major War', in *Security Dialogue* 21(3): 321–8.
Mutimer, D. (2007) 'Sovereign Contradictions: Maher Arar and the Indefinite Future', in E. Dauphinee & C. Masters (eds) *The Logics of Biopower and the War on Terror: Living, Dying, Surviving*. Basingstoke: Palgrave Macmillan, pp. 159–79.
Neumann, I. (1999) *Uses of the Other: "The East" in European Identity Formation*. London: University of Minnesota Press.
Noon, D. (2004) 'Operation Enduring Analogy: World War II, The War on Terror, and the Uses of Historical Memory', in *Rhetoric & Public Affairs* 7(3): 339–64.
Norris, C. (1990) 'Limited Think: How Not to Read Derrida', in *Diacritics* 20(1): 17–36.
Nowotny, H. (1992) 'Time and Social Theory: Towards a Social Theory of Time', in *Time & Society* 1(3): 421–54.
Nowotny, H. (1994) *Time: The Modern and Postmodern Experience*. Cambridge: Polity Press.
O'Brien, M. (1989) 'Periods', in F. Forman (ed.) *Taking Our Time: Feminist Perspectives on Temporality*. Oxford: Pergamon Press, pp. 11–18.
O'Malley, M. (1992) 'Time, Work and Task Orientation: A Critique of American Historiography', in *Time & Society* 1(3): 341–58.
Padelford, N. (1951) 'The Insecurity of States', in *The Western Political Quarterly* 4(3): 387–96.
Palmeri, F. (1999) 'History of Narrative Genres after Foucault', in *Configurations* 7(2): 267–77.
Parekh, B. (2002) 'Terrorism or Intercultural Dialogue', in K. Booth & T. Dunne (eds) *Worlds in Collision: Terror and the Future of Global Order*. Basingstoke: Palgrave Macmillan, pp. 270–83.

Patton, P. (2004) 'Politics', in J. Reynolds & J. Roffe (eds) *Understanding Derrida*. London: Continuum, pp. 26–36.
Pieterse, J. (2004) 'Neoliberal Empire', in *Theory, Culture and Society* 21(3): 119–40.
Post, J. (2005) 'When Hatred is Bred in the Bone: Psycho-Cultural Foundations of Contemporary Terrorism', in *Political Psychology* 26(4): 615–36.
Purvis, T. & Hunt, A. (1993) 'Discourse, Ideology, Discourse, Ideology, Discourse, Ideology ...', in *The British Journal of Sociology* 44(3): 473–99.
Rabinow, P. (2001) 'St. Francis at Ground Zero', in *Anthropological Quarterly* 75(1): 113–16.
Rackow, S. (2002) 'How the USA Patriot Act Will Permit Governmental Infringement upon the Privacy of Americans in the Name of "Intelligence" Investigations', in *University of Pennsylvania Law Review*: 150(5): 1651–96.
Rengger, N. (2002) 'On the Just War Tradition in the Twenty-First Century', in *International Affairs* 78(2): 353–63.
Richardson, B. (2006) 'Making Time: Narrative Temporality in Twentieth-Century Literature and Theory', in *Literature Compass* 3(3): 603–12.
Richardson, L. (2006) *What Terrorists Want: Understanding the Terrorist Threat*. London: John Murray.
Ricoeur, P. (1980) 'Narrative Time', in *Critical Inquiry* 7(1): 169–90.
Ricoeur, P. (1990) *Time and Narrative: Volume One*. London: University of Chicago Press.
Ricoeur, P. (1999) 'Imagination, Testimony and Trust: A Dialogue with Paul Ricoeur', in R. Kearney & M. Dooley (eds) *Questioning Ethics: Contemporary Debates in Philosophy*. London: Routledge, pp. 12–17.
Robbins, J. (2001) 'Secrecy and the Sense of an Ending: Narrative, Time, and Everyday Millenarianism in Papua New Guinea and in Christian Fundamentalism', in *Comparative Studies in Society and History* 43(3): 525–51.
Roberts, A. (2005) 'The "War on Terror" in Historical Perspective', in *Survival* 47(2): 101–30.
Said, E. (2003) *Orientalism* (4th Edn). London: Penguin.
Sassen, S. (2002) 'Governance Hotspots: Challenges We Must Confront in the Post-September 11 World', in K. Booth & T. Dunne (eds) *Worlds in Collision: Terror and the Future of Global Order*. Basingstoke: Palgrave Macmillan, pp. 313–24.
Schehr, L. (2006) 'Mr Malaprop, or, No President Left Behind', in *The South Atlantic Quarterly* 105(1): 137–52.
Schrecker, E. (2004) 'McCarthyism: Political Repression and the Fear of Communism', in *Social Research* 71(4): 1041–86.
Scraton, P. (ed.) (2002) *Beyond September 11: An Anthology of Dissent*. London: Pluto Press.
Silberstein, S. (2002) *War of Words: Language, Politics and 9/11*. London: Routledge.
Simon, S. & Benjamin, D. (2002) 'The Terror', in *Survival* 43(4): 5–18.
Simpson (2006) *9/11: The Culture of Commemoration*. Chicago: The University of Chicago Press.
Singer, P. (2004) *The President of Good and Evil: Taking George W. Bush Seriously*. London: Granta Books.
Smith, B. (1980) 'Narrative Versions, Narrative Theories', in *Critical Inquiry* 7(1): 213–36.

Smith, H. (2008) 'Ethics and Intelligence in the Age of terror', in A. Bellamy, R. Bleiker, S. Davies, & R. Devetak (eds) *Security and the War on Terror*. London: Routledge, pp. 156–76.

Sorokin, P. & Merton, R. (1937) 'Social Time: A Methodological and Functional Analysis', in *American Journal of Sociology* 42(5): 615–29.

Stanley, E. (2002) 'An Attack on Truth?', in P. Scraton (ed.) *Beyond September 11: An Anthology of Dissent*. London: Pluto Press, pp. 206–10.

Starr, C. (1966) 'Historical and Philosophical Time', in *History and Theory* 6(6): 24–35.

Stoddard, E. & Cornwell, G. (2002) 'Unity', in J. Collins & R. Glover (eds) *Collateral Language: A User's Guide to America's New War*. London: New York University Press, pp. 175–89.

Thelen, K. (1999) 'Historical Institutionalism in Comparative Politics', in *Annual Review of Political Science* 2(1): 369–404.

Thomas, L. (1996) 'Understanding Evil: American Slavery, the Holocaust and the Conquest of the American Indians', in *Ethics* 106(2): 424–48.

Thomassen, L. (2005) 'Antagonism, Hegemony and Ideology after Heterogeneity', in *Journal of Political Ideologies* 10(3): 289–309.

Thompson, E. (1967) 'Time, Work-Discipline and Industrial Capitalism', in *Past and Present* 38(1): 56–97.

Torfing, J. (1999) *New Theories of Discourse: Laclau, Mouffe and Žižek*. Oxford: Blackwell.

Torfing, J. (2005) 'Discourse Theory: Achievements, Arguments, and Challenges', in D. Howarth & J. Torfing (eds) *Discourse Theory in European Politics: Identity, Policy and Governance*. Basingstoke: Palgrave Macmillan, pp. 1–32.

Tracy, J. (2005) 'Bearing Witness to the Unspeakable: 9/11 and America's New Global Imperialism', in *The Journal of American Culture* 28(1): 85–99.

Tucker, D. (2001) 'What's New about the New Terrorism and how Dangerous is it?', in *Terrorism and Political Violence* 13(1): 1–14.

Turetzky, P. (1998) *Time* London: Routledge.

Turk, A. (2004) 'Sociology of Terrorism', in *Annual Review of Sociology* 30(1): 271–86.

Urry, J. (1994) 'Time, Leisure and Social Identity', in *Time & Society* 3(2): 131–49.

Van Alstyne, A. (2002) 'Freedom', in J. Collins & R. Glover (eds) *Collateral Language: A User's Guide to America's New War*. London: New York University Press, pp. 79–93.

Wagner-Pacifici, R. (1996) 'Memories in the Making: The Shapes of Things That Went', in *Qualitative Sociology* 19(3): 301–21.

Waldron, J. (2003) 'Security and Liberty: The Image of Balance', in *Journal of Political Philosophy* 11(2): 191–210.

Watson, M. & Hay, C. (2003) 'The Discourse of Globalization and the Logic of No Alternative: Rendering the Contingent Necessary in the Political Economy of New Labour', in *Policy & Politics* 31(3): 289–305.

Weber, C. (2007). 'Securitizing the Unconscious: The Bush Doctrine of Preemption and *Minority Report*', in E. Dauphinee & C. Masters (eds) *The Logics of Biopower and the War on Terror: Living, Dying, Surviving*. Basingstoke: Palgrave Macmillan, pp. 109–27.

White, H. (1980) 'The Value of Narrativity in the Representation of Reality', in *Critical Inquiry* 7(1): 5–27.

White, H. (1990) *The Content of the Form: Narrative Discourse and Historical Representation*. London: John Hopkins University Press.
Williams, P. (2002) 'Peace, Poetry and Pentagonese', in K. Booth & T. Dunne (eds) *Worlds in Collision: Terror and the Future of Global Order.* Basingstoke: Palgrave Macmillan, pp. 336–47.
Williams, P. (2008) 'Security Studies, 9/11 and the Long War', in A. Bellamy, R. Bleiker, S. Davies & R. Devetak (eds) *Security and the War on Terror.* London: Routledge, pp. 9–24.
Zerubavel, E. (1977) 'The French Republican Calendar: A Case Study in the Sociology of Time', in *American Sociological Review* 42(6): 868–77.
Žižek, S. (2002) *Welcome to the Desert of the Real!: Five Essays on September 11 and Related Dates*. London: Verso.
Zulaika, J. & Douglass, W. (1996) *Terror and Taboo: The Follies, Fables and Faces of Terrorism*. London: Routledge.

List of Sources

Ashcroft, J. (11/9/2001) 'Remarks of Attorney General John Ashcroft'. Available at: http://www.usdoj.gov/archive/ag/speeches/2001/agcrisisremarks.htm Accessed: 14.4.2006.

Ashcroft, J. (13/9/2001) 'Prepared Remarks: Attorney General John Ashcroft. Press Availability'. Available at: http://www.usdoj.gov/archive/ag/speeches/2001/0913pressconference.htm Accessed: 14.4.2006.

Ashcroft, J. (18/9/2001) 'America's New War, Justice Department Briefing'. Available at: http://transcripts.cnn.com/TRANSCRIPTS/0109/18/se.18.html Accessed: 22.7.2006.

Ashcroft, J. & Mueller, R. (20/9/2001) 'Attorney General Remarks, Pennsylvania Crash Site'. Available at: http://www.usdoj.gov/archive/ag/speeches/2001/agcrisisremarks9_20.htm Accessed: 11.9.2003.

Ashcroft, J., Mueller, R., Giuliani, R. & Pataki, G. (21/9/2001) 'Remarks from Press Conference at WTC, New York'. Available at: http://www.usdoj.gov/archive/ag/speeches/2001/agcrisisremarks9_21pc.htm Accessed: 17.4.2003.

Ashcroft, J. & Mueller, R. (21/9/2001) 'Attorney General and FBI Director Remarks: Tour of the World Trade Center'. Available at: http://www.usdoj.gov/archive/ag/speeches/2001/agcrisisremarks9_21.htm Accessed: 14.4.2006.

Ashcroft, J. (25/9/2001) 'Attorney General John Ashcroft Testimony before the Senate Committee on the Judiciary'. Available at: http://www.usdoj.gov/archive/ag/testimony/2001/0925AttorneyGeneralJohnAshcroftTestimonybeforetheSenateCommitteeontheJudiciary.htm Accessed: 11.9.2004.

Ashcroft, J. & MacAulay, L. (2/10/2001) 'John Ashcroft Holds Media Availability with Canadian Solicitor General Lawrence MacAulay'. Available at: http://www.usdoj.gov/archive/ag/speeches/2001/agcrisisremarks10_2.htm Accessed: 14.4.2006.

Ashcroft, J. (25/10/2001) 'Attorney General John Ashcroft Prepared Remarks for the US Mayors Conference'. Available at: http://www.usdoj.gov/archive/ag/speeches/2001/agcrisisremarks10_25.htm Accessed: 11.9.2004.

Ashcroft, J. (14/11/2001) 'Attorney General John Ashcroft and INS Commissioner Ziglar Announce INS Restructuring Plan'. Available at: http://www.usdoj.gov/archive/ag/speeches/2001/agcrisisremarks11_14.htm Accessed: 11.9.2004.

Ashcroft, J. (26/11/2001) 'Attorney General Ashcroft Announces the Appointment of the Special Master to Administer the September 11 Victim Compensation Fund'. Available at: http://www.usdoj.gov/archive/ag/speeches/2001/agcrisisremarks11_26.htm Accessed: 11.9.2004.

Ashcroft, J. (6/12/2001) 'Testimony of Attorney General John Ashcroft, Senate Committee on the Judiciary'. Available at: http://www.usdoj.gov/archive/ag/testimony/2001/1206transcriptsenatejudiciarycommittee.htm Accessed: 11.9.2004.

Ashcroft, J. (24/1/2002) 'Prepared Remarks of Attorney General John Ashcroft, Hispanic American Police Command Officers Association'. Available at:

http://www.usdoj.gov/archive/ag/speeches/2002/012402preparedremarks.htm Accessed: 11.9.2004.

Ashcroft, J. (28/1/2002) 'Prepared Remarks of Attorney General John Ashcroft, Organization of American States, Inter-American Committee Against Terrorism'. Available at: http://www.usdoj.gov/archive/ag/speeches/2002/012802agpreparedremarks.htm Accessed: 11.9.2004.

Ashcroft, J. (19/2/2002) 'Prepared Remarks of Attorney General John Ashcroft, National Religious Broadcasters Convention, Nashville, Tennessee'. Available at: http://www.usdoj.gov/archive/ag/speeches/2002/021902religiousbroadcasters.htm Accessed: 11.9.2004.

Ashcroft, J. (26/2/2002) 'Statement of John Ashcroft, Attorney General of the United States before the United States Senate Committee on Appropriations, Subcommittee on the Departments of Commerce, Justice, and State, the Judiciary, and Related Agencies'. Available at: http://www.usdoj.gov/archive/ag/testimony/2002/FY2003AG_WrittenStatement-Senate.htm Accessed: 11.9.2004.

Ashcroft, J. (18/5/2002) 'Prepared Remarks of Attorney General John Ashcroft, Commencement Address for the University of Missouri, Columbia School of Law, Jesse Auditorium'. Available at: http://www.usdoj.gov/archive/ag/speeches/2002/051802commencement_columbia.htm Accessed: 11.9.2004.

Ashcroft, J. (25/7/2002) 'Statement of John Ashcroft, Attorney General of the United States before the Committee on the Judiciary, United States Senate'. Available at: http://www.usdoj.gov/ag/testimony/2002/072502agtestimony.htm Accessed: 11.9.2004.

Ashcroft, J. (7/8/2002) 'Remarks of Attorney General John Ashcroft, Eighth Circuit Judges Conference, Duluth, Minnesota'. Available at: http://www.usdoj.gov/archive/ag/speeches/2002/080702eighthcircuitjudgesagremarks.htm Accessed: 3.1.2006.

Ashcroft, J. & Ridge, T. (10/9/2002) 'Director Ridge, Attorney General Ashcroft Discuss Threat Level, The Justice Department'. Available at: http://www.dhs.gov/xnews/speeches/speech_0029.shtm Accessed: 3.1.2006.

Ashcroft, J. (14/9/2002) 'Remarks of Attorney General John Ashcroft, European Union Luncheon'. Available at: http://www.usdoj.gov/archive/ag/speeches/2002/091402agremarkseuministers.htm Accessed: 11.9.2004.

Ashcroft, J. (16/9/2002) 'Remarks of Attorney General John Ashcroft, Meeting in Belgium'. Available at: http://www.usdoj.gov/archive/ag/speeches/2002/091602agremarksbelgium.htm Accessed: 11.9.2004.

Ashcroft, J. (1/10/2002) 'Remarks of Attorney General John Ashcroft, US Attorneys Conference, New York City'. Available at: http://www.usdoj.gov/archive/ag/speeches/2002/100102agremarkstousattorneysconference.htm Accessed: 11.9.2004.

Ashcroft, J. (7/10/2002) 'Prepared Remarks of Attorney General John Ashcroft, International Association of Chiefs of Police Conference, Minneapolis, Maine'. Available at: http://www.usdoj.gov/archive/ag/speeches/2002/100702chiefsofpolicemn1.htm Accessed: 11.9.2004.

Ashcroft, J. (9/9/2003) 'Prepared Remarks of Attorney General John Ashcroft, New York City'. Available at: http://www.usdoj.gov/archive/ag/speeches/2003/090903nycremarks.htm Accessed: 11.9.2004.

Ashcroft, J. (7/11/2003) 'Prepared Remarks of Attorney General John Ashcroft, New York, New York'. Available at: http://www.usdoj.gov/archive/ag/speeches/2003/agantidefamation.htm Accessed: 16.3.2006.

Bush, G. (11/9/2001a) 'Remarks by the President after Two Planes Crash into World Trade Center, Emma Booker Elementary School, Sarasota, Florida'. Available at: http://www.whitehouse.gov/news/releases/2001/09/20010911.html Accessed: 11.9.2004.

Bush, G. (11/9/2001b) 'Remarks by the President upon Arrival at Barksdale Air Force Base, Barksdale Air Force Base, Louisiana'. Available at: http://www.whitehouse.gov/news/releases/2001/09/20010911-1.html Accessed: 11.9.2004.

Bush, G. (11/9/2001c) 'Statement by the President in His Address to the Nation'. Available at: http://www.whitehouse.gov/news/releases/2001/09/20010911-16.html Accessed: 11.9.2004.

Bush, G. (12/9/2001a) 'Presidential Memo to Executive Departments and Agencies'. Available at: http://www.whitehouse.gov/news/releases/2001/09/20010912-13.html Accessed: 11.9.2004.

Bush, G. (12/9/2001b) 'Honoring the Victims of the Incidents on Tuesday, 11 September, 2001: By the President of the United States of America. A Proclamation'. Available at: http://www.whitehouse.gov/news/releases/2001/09/20010912-1.html Accessed: 11.9.2004.

Bush, G. (12/9/2001c) 'Remarks by the President in Photo Opportunity with the National Security Team, The Cabinet Room'. Available at: http://www.whitehouse.gov/news/releases/2001/09/20010912-4.html Accessed: 11.9.2004.

Bush, G. (13/9/2001a) 'Remarks by the President to the Travel Pool after Visiting Washington Hospital Center, Washington Hospital Center'. Available at: http://www.whitehouse.gov/news/releases/2001/09/20010913-5.html Accessed: 11.9.2004.

Bush, G. (13/9/2001b) 'National Day of Prayer of and Remembrance for the Victims of the Terrorist Attacks on 11 September, 2001: By the President of the United States of America. A Proclamation'. Available at: http://www.whitehouse.gov/news/releases/2001/09/20010913-7.html Accessed: 11.9.2004.

Bush, G., Pataki, G. & Guiliani, R. (13/9/2001) 'President Pledges Assistance for New York in Phone Call with Pataki, Guiliani'. Available at: http://www.whitehouse.gov/news/releases/2001/09/20010913-4.html Accessed: 11.9.2004.

Bush, G. (14/9/2001a) 'President's Remarks at National Day of Prayer and Remembrance, The National Cathedral, Washington, DC'. Available at: http://www.whitehouse.gov/news/releases/2001/09/20010914-2.html Accessed: 11.9.2004.

Bush, G. (14/9/2001b) 'President Extends Order for Flags at Half-Staff: Amending Proclamation 7461, Display of the Flag at Half-Staff as a Mark of Respect for the Victims of the Incidents on Tuesday, 11 September, 2001: By the President of the United States of America. A Proclamation'. Available at: http://www.whitehouse.gov/news/releases/2001/09/20010914-7.html Accessed: 11.9.2004.

Bush, G. (15/9/2001) 'Radio Address of the President to the Nation'. Available at: http://www.whitehouse.gov/news/releases/2001/09/20010915.html Accessed: 11.9.2004.

Bush, G., Powell, C. & Ashcroft, J. (15/9/2001) 'President Urges Readiness and Patience, Remarks by the President, Secretary of State Colin Powell and Attorney General John Ashcroft, Camp David, Thurmont, Maryland'. Available at: http://www.whitehouse.gov/news/releases/2001/09/20010915-4.html Accessed: 11.9.2004.

Bush, G. (16/9/2001) 'Remarks by the President upon Arrival, The South Lawn'. Available at: http://www.whitehouse.gov/news/releases/2001/09/20010916-2.html Accessed: 11.9.2004.

Bush, G. (17/9/2001a) 'Guards and Reserves "Define Spirit of America", Remarks by the President to Employees at the Pentagon, The Pentagon'. Available at: http://www.whitehouse.gov/news/releases/2001/09/20010917-3.html Accessed: 11/9/2003.

Bush, G. (17/9/2001b) '"Islam is Peace" Says President, Remarks by the President at Islamic Center of Washington DC, Washington, DC'. Available at: http://www.whitehouse.gov/news/releases/2001/09/20010917-11.html Accessed: 11.9.2004.

Bush, G. (19/9/2001) 'Remarks by the President at Photo Opportunity with House and Senate leadership'. Available at: http://www.whitehouse.gov/news/releases/2001/09/20010919-8.html Accessed: 11.9.2004.

Bush, G. & Megawati, S. (19/9/2001) 'President Building Worldwide Campaign against Terrorism'. Available at: http://www.whitehouse.gov/news/releases/2001/09/20010919-1.html Accessed: 11.9.2004.

Bush, G. (20/9/2001) 'Address to a Joint Session of Congress, and the American People, United States Capitol, Washington DC'. Available at: http://www.whitehouse.gov/news/releases/2001/09/20010920-8.html Accessed: 11.9.2004.

Bush, G., O'Neill, P. & Powell, C. (24/9/2001) 'President Freezes Terrorists' Assets, Remarks by the President, Secretary of the Treasury O'Neill and Secretary of State Powell on Executive Order, The Rose Garden'. Available at: http://www.whitehouse.gov/news/releases/2001/09/20010924-4.html Accessed: 11.9.2004.

Bush, G. (25/9/2001) 'President: FBI Needs Tools to Track Down Terrorists, Remarks by the President to Employees at the Federal Bureau of Investigation, FBI Headquarters, Washington, DC'. Available at: http://www.whitehouse.gov/news/releases/2001/09/20010925-5.html Accessed: 11.9.2004.

Bush, G. (26/9/2001a) 'President Meets with Muslim Leaders: 26 September 2001'. Available at: http://www.whitehouse.gov/news/releases/2001/09/20010926-8.html Accessed: 11.9.2004.

Bush, G. (26/9/2001b) 'President Thanks CIA, Remarks by the President to Employees of the Central Intelligence Agency, CIA Headquarters, Langley, Virginia'. Available at: http://www.whitehouse.gov/news/releases/2001/09/20010926-3.html Accessed: 11.9.2004.

Bush, G. (27/9/2001) 'At O'Hare, President Says "Get on Board", Remarks by the President to Airline Employees, O'Hare International Airport, Chicago, Illinois'. Available at: http://www.whitehouse.gov/news/releases/2001/09/20010927-1.html Accessed: 11.9.2004.

Bush, G. (29/9/2001) 'Radio Address of the President to the Nation'. Available at: http://www.whitehouse.gov/news/releases/2001/09/20010929.html Accessed: 11.9.2004.

Bush, G. (1/10/2001) 'President: "We're Making Progress", Remarks by the President to Employees of FEMA, FEMA Headquarters, Washington, DC'. Available at: http://www.whitehouse.gov/news/releases/2001/10/20011001-6.html Accessed: 11.9.2004.

Bush, G. (2/10/2001a) 'President Opens Reagan Airport'. Available at: http://www.whitehouse.gov/news/releases/2001/10/20011002-8.html Accessed: 11.9.2004.

Bush, G. (2/10/2001b) 'President Meets with Congressional Leaders, The Oval Office'. Available at: http://www.whitehouse.gov/news/releases/2001/10/20011002-1.html Accessed: 11.9.2004.

Bush, G. (3/10/2001) 'President Discusses Economic Recovery in New York City, Freedom Hall, New York, New York'. Available at: http://www.whitehouse.gov/news/releases/2001/10/20011003-4.html Accessed: 11.9.2004.

Bush, G. (4/10/2001a) 'President Unveils Back to Work Plan, Department of Labor, Washington, DC'. Available at: http://www.whitehouse.gov/news/releases/2001/10/20011004-8.html Accessed: 11.9.2004.

Bush, G. (4/10/2001b) 'President Directs Humanitarian Aid to Afghanistan, Remarks by the President to State Department Employees, U.S. Department of State, Washington, DC'. Available at: http://www.whitehouse.gov/news/releases/2001/10/20011004.html Accessed: 11.9.2004.

Bush, G. (7/10/2001) 'President Speaks at 20th Annual National Fallen Firefighters Memorial Tribute'. Available at: http://www.whitehouse.gov/news/releases/2001/10/20011007-2.html Accessed: 11.9.2004.

Bush, G. (8/10/2001) 'Remarks by the President in Signing of the Columbus Day Proclamation'. Available at: http://www.whitehouse.gov/news/releases/2001/10/20011008-8.html Accessed: 11.9.2004.

Bush, G. & Schroeder, G. (9/10/2001) 'German Leader Reiterates Solidarity with U.S., Remarks by the President and Chancellor Schroeder of Germany in Press Availability, The Colonnade'. Available at: http://www.whitehouse.gov/news/releases/2001/10/20011009-13.html Accessed: 11.9.2004.

Bush, G. (10/10/2001a) 'General Pulaski Memorial Day Proclamation: By the President of the United States. A Proclamation'. Available at: http://www.whitehouse.gov/news/releases/2001/10/20011010-4.html Accessed: 11.9.2004.

Bush, G. (10/10/2001b) 'President Unveils "Most Wanted" Terrorists, Remarks by the President During Announcement at the Federal Bureau of Investigation, FBI Headquarters, Washington, DC'. Available at: http://www.whitehouse.gov/news/releases/2001/10/20011010-3.html Accessed: 11.9.2004.

Bush, G. (11/10/2001a): 'President Pays Tribute at Pentagon Memorial, Remarks by President at the Department of Defense Service of Remembrance, The Pentagon, Arlington, Virginia'. Available at: http://www.whitehouse.gov/news/releases/2001/10/20011011-1.html Accessed: 11.9.2004.

Bush, G. (11/10/2001b) 'President Holds Prime Time News Conference, The East Room'. Available at: http://www.whitehouse.gov/news/releases/2001/10/20011011-7.html Accessed: 30.7.2004.

Bush, G. (11/10/2001c) 'President Announces "America's Fund for Afghan Children", The East Room'. Available at: http://www.whitehouse.gov/news/releases/2001/10/20011011-8.html Accessed 11.9.2004.

Bush, G. (12/10/2001) 'President's Remarks during Hispanic Heritage Month Event'. Available at: http://www.whitehouse.gov/news/releases/2001/10/20011012-6.html Accessed: 11.9.2004.

Bush, G. (13/10/2001) 'Radio Address of the President to the Nation'. Available at: http://www.whitehouse.gov/news/releases/2001/10/20011013.html Accessed: 11.9.2004.

Bush, G. (14/10/2001) 'President Discusses Medicare to the American Society of Anaesthesiologists'. Available at: http://www.whitehouse.gov/news/releases/2001/10/20011014.html Accessed: 11.9.2004.
Bush, G. (15/10/2001) 'President Honors Public Servants'. Available at: http://www.whitehouse.gov/news/releases/2001/10/20011015-8.html Accessed: 11.9.2004.
Bush, G. (17/10/2001) 'President Outlines War Effort, Remarks by the President at the California Business Association Breakfast, Sacramento, California'. Available at: http://www.whitehouse.gov/news/releases/2001/10/20011017-15.html Accessed: 11.9.2004.
Bush, G. (24/10/2001) 'President Discusses Stronger Economy and Homeland Defense, Dixie Printing Company, Glen Burnie, Maryland'. Available at: http://www.whitehouse.gov/news/releases/2001/10/20011024-2.html Accessed: 11.9.2004.
Bush, G. (26/10/2001) 'President Signs Anti-Terrorism Bill, Remarks by the President at Signing of the Patriot Act, Anti-Terrorism Legislation, The East Room'. Available at: http://www.whitehouse.gov/news/releases/2001/10/20011026-5.html Accessed: 11.9.2004.
Bush, G. (29/10/2001) 'U.S. Africa Strengthen Counter-Terrorism and Economic Ties, Remarks by the President to the African Growth and Opportunity Forum, Department of State, Washington, DC'. Available at: http://www.whitehouse.gov/news/releases/2001/10/20011029-2.html Accessed: 11.9.2004.
Bush, G. (2/11/2001) 'President Salutes USO in White House Ceremony, Remarks by the President at Reception Honoring United Service Organization, The East Room'. Available at: http://www.whitehouse.gov/news/releases/2001/11/20011102-14.html Accessed: 11.9.2004.
Bush, G. (6/11/2001) 'President Bush: "No Nation Can be Neutral in This Conflict", Remarks to the President to the Warsaw Conference for Combating Terrorism'. Available at: http://www.whitehouse.gov/news/releases/2001/11/20011106-2.html Accessed: 11.9.2004.
Bush, G. (8/11/2001) 'President Discusses War on Terrorism, in Address to the Nation, World Congress Center, Atlanta, Georgia'. Available at: http://www.whitehouse.gov/news/releases/2001/11/20011108-13.html Accessed: 11.9.2001.
Bush, G. (11/11/2001) 'President Speaks at Veterans Day Prayer Breakfast, Park Avenue Seventh Regiment Armory, New York, New York'. Available at: http://www.whitehouse.gov/news/releases/2001/11/20011111.html Accessed: 11.9.2001.
Bush, G. (19/11/2001) 'President Discusses War, Humanitarian Efforts, Remarks by the President in Photo Opportunity with the Cabinet, The Cabinet Room'. Available at: http://www.whitehouse.gov/news/releases/2001/11/20011119-12.html Accessed: 11.9.2004.
Bush, G. & Arroyo, G. (20/11/2001) 'Philippine President Reiterates Support, Remarks by President Bush and President Arroyo of the Philippines in a Photo Opportunity, The Oval Office'. Available at: http://www.whitehouse.gov/news/releases/2001/11/20011120-11.html Accessed: 11.9.2004.
Bush, G. (21/11/2001) 'President Shares Thanksgiving Meal with Troops, Remarks by the President to Troops and Families at Fort Campbell, Fort Campbell,

Kentucky'. Available at: http://www.whitehouse.gov/news/releases/2001/11/20011121-3.html Accessed: 11.9.2004.
Bush, G. (29/11/2001) 'President Says U.S. Attorneys on Front Line in War, Remarks by the President to U.S. Attorneys Conference, Presidential Hall, Dwight David Eisenhower Office Building'. Available at: http://www.whitehouse.gov/news/releases/2001/11/20011129-12.html Accessed: 11.9.2004.
Bush, G. (4/12/2001) 'President Meets with Displaced Workers in Town Hall Meeting, Orange County Convention Center, Orlando, Florida'. Available at: http://www.whitehouse.gov/news/releases/2001/12/20011204-17.html Accessed: 11.9.2004.
Bush, G. & Bondevik, K. (5/12/2001) 'President Asks Global Leaders to Play Anthems on September 11, Remarks by the President and Prime Minister Kjell Magne Bondevik of Norway in Photo Opportunity, The Oval Office'. Available at: http://www.whitehouse.gov/news/releases/2001/12/20011205-11.html Accessed: 11.9.2004.
Bush, G. (7/12/2001) 'We're Fighting to Win – and Win We Will: Remarks by the President on the USS Enterprise on Pearl Harbor Day'. Available at: http://www.whitehouse.gov/news/releases/2001/12/20011207.html Accessed: 11.9.2003.
Bush, G. (11/12/2001) 'President: The World Will Always Remember September 11'. Available at: http://www.whitehouse.gov/news/releases/2001/12/20011211-1.html Accessed: 11.9.2004.
Bush, G. & Shinawatra, T. (14/12/2001) 'Prime Minister of Thailand Reiterates Full Support to US, Remarks by the President and Prime Minister Thaksin Shinawatra of Thailand, in Photo Opportunity, The Oval Office'. Available at: http://www.whitehouse.gov/news/releases/2001/12/20011214-5.html Accessed: 11.9.2004.
Bush, G. & Franks, T. (28/12/2001) 'President, General Franks Discuss War Effort, Remarks by the President and General Tommy Franks in Press Availability with the Press Travel Pool, The Prairie Chapel Ranch, Crawford, Texas'. Available at: http://www.whitehouse.gov/news/releases/2001/12/20011228-1.html Accessed: 11.9.2004.
Bush, G. (5/1/2002) 'President Holds Town Hall Forum on Economy in California, Ontario Convention Center, Ontario, California'. Available at: http://www.whitehouse.gov/news/releases/2002/01/20020105-3.html Accessed: 11.9.2004.
Bush, G. (24/1/2002) 'President Announces Substantial Increases in Homeland Security Budget, Remarks by the President to U.S. Mayors and County Officials, The East Room'. Available at: http://www.whitehouse.gov/news/releases/2002/01/20020124-1.html Accessed: 11.9.2004.
Bush, G. (29/1/2002) 'President Delivers State of the Union Address'. Available at: http://www.whitehouse.gov/news/releases/2002/01/20020129-11.html Accessed: 11.9.2004.
Bush, G. (5/2/2002) 'President Increases Funding for Bioterrorism by 319 Percent, Masonic Temple, University of Pittsburgh, Pittsburgh, Pennsylvania'. Available at: http://www.whitehouse.gov/news/releases/2002/02/20020205-4.html Accessed: 11.9.2004.
Bush, G. (6/2/2002) 'Remarks by the President at Friends of Governor Pataki Reception, The Sheraton, New York'. Available at: http://www.whitehouse.gov/news/releases/2002/02/20020206-8.html Accessed: 11.9.2004.

Bush, G. (7/2/2002) 'President's Remarks at National Prayer Breakfast, Remarks by the President at National Prayer Breakfast, Washington Hilton Hotel, Washington, DC'. Available at: http://www.whitehouse.gov/news/releases/2002/02/20020207-1.html Accessed: 11.9.2004.

Bush, G. (11/2/2002) 'Remarks by the President at Scott Mccallum for Governor Reception, The Pfister Hotel, Milwaukee, Wisconsin'. Available at: http://www.whitehouse.gov/news/releases/2002/02/20020211-9.html Accessed: 11.9.2004.

Bush, G. (11/3/2002a) 'President Thanks World Coalition for Anti-terrorism Efforts'. Available at: http://www.whitehouse.gov/news/releases/2002/03/20020311-1.html Accessed: 11.9.2004.

Bush, G. (11/3/2002b) 'President Unveils September 11 Postage Stamp, Remarks by the President in Unveiling of September 11th United States Postage Stamp, The Oval Office'. Available at: http://www.whitehouse.gov/news/releases/2002/03/20020311-4.html Accessed: 11.9.2004.

Bush, G. (25/3/2002) 'President Recognizes Greek Independence Day, Remarks by the President at Greek Independence Day Ceremony, Presidential Hall, Eisenhower Executive Office Building'. Available at: http://www.whitehouse.gov/news/releases/2002/03/20020325-1.html Accessed: 11.9.2004.

Bush, G. (30/3/2002) 'President Promotes Peace in Radio Address, Radio Address by the President to the Nation'. Available at: http://www.whitehouse.gov/news/releases/2002/03/20020330.html Accessed: 11.9.2004.

Bush, G. (3/4/2002) 'Remarks by the President at Fisher for Governor Reception, Four Seasons Hotel, Philadelphia, Pennsylvania'. Available at: http://www.whitehouse.gov/news/releases/2002/04/20020403-1.html Accessed: 11.9.2004.

Bush, G. (17/4/2002) 'President Outlines War Effort, Remarks by the President to the George C. Marshall ROTC Award Seminar on National Security, Cameron Hall, Virginia Military Institute, Lexington, Virginia'. Available at: http://www.whitehouse.gov/news/releases/2002/04/20020417-1.html Accessed: 11.9.2004.

Bush, G. (18/4/2002) 'National Crime Victims' Rights Week, 2002, By the President of the United States. A Proclamation' Available at: http://www.whitehouse.gov/news/releases/2002/04/20020418-20.html Accessed: 11.9.2004.

Bush, G. (30/4/2002) 'President Promotes Compassionate Conservatism, Parkside Hall, San Jose, California'. Available at: http://www.whitehouse.gov/news/releases/2002/04/20020430-5.html Accessed: 11.9.2004.

Bush, G. (1/5/2002) 'Remarks by the President at Simon for Governor Luncheon, Santa Clara Convention Center, Santa Clara, California'. Available at: http://www.whitehouse.gov/news/releases/2002/05/20020501-2.html Accessed: 11.9.2004.

Bush, G. (4/5/2002) 'Radio Address by the President to the Nation'. Available at: http://www.whitehouse.gov/news/releases/2002/05/20020504.html Accessed: 11.9.2004.

Bush, G. (9/5/2002) 'Statement by the President'. Available at: http://www.whitehouse.gov/news/releases/2002/05/20020509-10.html Accessed: 11.9.2004.

Bush, G. (17/5/2002) 'President Discusses Response to September 11 Attacks, Remarks by the President at Presentation of the Commander-In-Chief's Trophy, The Rose Garden'. Available at: http://www.whitehouse.gov/news/releases/2002/05/20020517-1.html Accessed: 3.1.2006.

Bush, G. (23/5/2002) 'President Bush Thanks Germany for Support against Terror, Remarks by the President to a Special Session of the German Bundestag,

The Bundestag, Berlin, Germany'. Available at: http://www.whitehouse.gov/news/releases/2002/05/20020523-2.html Accessed: 3.1.2006.
Bush, G. (6/6/2002) 'Remarks by the President in Address to the Nation, The Cross Hall'. Available at: http://www.dhs.gov/xnews/speeches/speech_0048.shtm Accessed: 3.1.2006.
Bush, G. (7/6/2002) 'President Meets with Congressional Leaders on Homeland Security, The Cabinet Room'. Available at: http://www.dhs.gov/xnews/speeches/speech_0047.shtm Accessed: 3.1.2006.
Bush, G. (11/6/2002) 'President Discusses Nation's Critical Infrastructure, Oak Park High School, Kansas City, Missouri'. Available at: http://www.dhs.gov/xnews/speeches/speech_0042.shtm Accessed: 3.1.2006.
Bush, G. (13/6/2002) 'Statement by the President'. Available at: http://www.whitehouse.gov/news/releases/2002/06/20020613-9.html Accessed: 3.1.2006.
Bush, G. (14/6/2002) 'Remarks by the President in Texans for Rick Perry Reception, Hyatt Regency Hotel, Houston, Texas'. Available at: http://www.whitehouse.gov/news/releases/2002/06/20020614-8.html Accessed: 3.1.2006.
Bush, G. (18/6/2002) 'Homeland Security Proposal Delivered to Congress'. Available at: http://www.dhs.gov/xnews/speeches/speech_0039.shtm Accessed: 3.1.2006.
Bush, G. (24/6/2002) 'Remarks by the President at Ferguson for Congress Luncheon, Sheraton Newark Hotel, Newark, Jersey'. Available at: http://www.whitehouse.gov/news/releases/2002/06/20020624-6.html Accessed: 3.1.2006.
Bush, G. (6/7/2002) 'President's Radio Address'. Available at: http://www.whitehouse.gov/news/releases/2002/07/20020706.html Accessed: 3.1.2006.
Bush, G. (10/7/2002) 'Remarks by the President to Federal Employees on Homeland Security, DAR Constitution Hall, Washington, DC'. Available at: http://www.dhs.gov/xnews/speeches/speech_0070.shtm Accessed: 3.1.2006.
Bush, G. (17/7/2002) 'Captive Nations Week, 2002, by The President of the United States: A Proclamation'. Available at: http://www.whitehouse.gov/news/releases/2002/07/20020717-10.html Accessed: 3.1.2006.
Bush, G. (26/7/2002) 'Managerial Flexibility Key to Successful Homeland Security Dept, Dwight D. Eisenhower Executive Office Building'. Available at: http://www.dhs.gov/xnews/speeches/speech_0032.shtm Accessed: 3.1.2006.
Bush, G. (14/8/2002) 'President's Priorities: National Security, Homeland Security, Economic Security, Remarks by the President at Milwaukee, Wisconsin Welcome, Klotsche Center'. Available at: http://www.dhs.gov/xnews/speeches/speech_0031.shtm Accessed: 3.1.2006.
Bush, G. (15/8/2002) 'President Talks Homeland/Economic Security at Mt. Rushmore, Remarks by the President on Homeland Security and the Budget, Mount Rushmore, South Dakota'. Available at: http://www.dhs.gov/xnews/speeches/speech_0030.shtm Accessed: 3.1.2006.
Bush, G. (4/9/2002) 'Patriot Day, 2002: By the President of the United States of America. A Declaration'. Available at: http://www.whitehouse.gov/news/releases/2002/09/20020904-10.html Accessed: 3.1.2006.
Bush, G. (5/9/2002) 'Remarks by the President at Chris Chocola for Congress and Indiana Victory 2002 Finance Dinner, Century Center, South Bend, Indiana'. Available at: http://www.whitehouse.gov/news/releases/2002/09/20020905-7.html Accessed: 3.1.2006.

Bush, G. (9/9/2002) 'President Bush, PM Chretien Announce Progress in Smart Border Plan'. Available at: http://www.dhs.gov/xnews/releases/press_release_0023.shtm Accessed: 3.1.2006.

Bush, G. (11/9/2002a) 'President's Remarks at the Pentagon, Remarks by the President in Observance of September 11th, The Pentagon'. Available at: http://www.whitehouse.gov/news/releases/2002/09/20020911.html Accessed: 11.9.2004.

Bush, G. (11/9/2002b) 'President's Remarks to the Nation, Ellis Island, New York'. Available at: http://www.whitehouse.gov/news/releases/2002/09/20020911-3.html Accessed: 11.9.2004.

Bush, G. (28/9/2002) 'President Bush Pushes for Homeland Security Bill, Remarks by the President at Matt Salmon for Governor and Rick Renzi for Congress Dinner, Phoenix Civic Plaza, Phoenix, Arizona'. Available at: http://www.dhs.gov/xnews/speeches/speech_0024.shtm Accessed: 3.1.2006.

Bush, G. (12/11/2002) 'President Bush Pushes for Homeland Security Department, Remarks by the President at District of Columbia Metropolitan Police Operations Center, Washington, DC'. Available at: http://www.dhs.gov/xnews/speeches/speech_0020.shtm Accessed: 3.1.2006.

Bush, G. (13/11/2002) 'President Discusses Homeland Security/Economy with Cabinet, Remarks with the President after Meeting with the Cabinet, The Cabinet Room'. Available at: http://www.dhs.gov/xnews/speeches/speech_0019.shtm Accessed: 3.1.2006.

Bush, G. (23/11/2002) 'President Bush Welcomes Romania to Nato, Remarks by the President to the Citizens of Romania, Piati, Revolutiei, Bucharest, Romania'. Available at: http://www.whitehouse.gov/news/releases/2002/11/20021123-7.html Accessed: 3.1.2006.

Bush, G. (6/12/2002) 'National Pearl Harbor Remembrance Day 2002, by the President of the United States of America: A Proclamation'. Available at: http://www.whitehouse.gov/news/releases/2002/12/20021206-5.html Accessed: 3.1.2006.

Bush, G. (28/1/2003) 'SOTU Excerpts on Defending Peace and Security at Home'. Available at: http://www.whitehouse.gov/news/releases/2003/01/20030128-22.html Accessed: 3.1.2006.

Bush, G. (26/3/2003) 'President Rallies Troops at MacDill Air Force Base in Tampa, Remarks by the President to Socom and Centcom Community, MacDill Air Force Base, Tampa, Florida'. Available at: http://www.whitehouse.gov/news/releases/2003/03/20030326-4.html Accessed: 3.1.2006.

Bush, G. (3/4/2003) 'President Discusses Operation Iraqi Freedom at Camp Lejeune, Remarks by the President to the Military Personnel and Their Families, Marine Corps Base Camp, Lejeune, Camp Lejeune, North Carolina'. Available at: http://www.whitehouse.gov/news/releases/2003/04/20030403-3.html Accessed: 3.1.2006.

Bush, G. (12/4/2003) 'Operation Iraqi Freedom: President's Radio Address'. Available at: http://www.whitehouse.gov/news/releases/2003/04/20030412.html Accessed: 3.1.2006.

Bush, G. (6/11/2003) 'President Bush Discusses Freedom in Iraq and Middle East, Remarks by the President at the 20th Anniversary of the National Endowment for Democracy, United States Chamber of Commerce, Washington, DC'.

Available at: http://www.whitehouse.gov/news/releases/2003/11/20031106-2.html Accessed: 3.1.2006.

Bush, G. (20/1/2004) 'State of the Union Address, United States Captiol, Washington, DC'. Available at: http://www.whitehouse.gov/news/releases/2004/01/20040120-7.html Accessed: 3.1.2006.

Bush, G. (11/6/2004) 'President Bush's Eulogy at Funeral Service for President Reagan, Remarks by the President in Eulogy at National Funeral Service for Former President Ronald Wilson Reagan, The National Cathedral, Washington, DC'. Available at: http://www.whitehouse.gov/news/releases/2004/06/20040611-2.html Accessed: 3.1.2006.

Bush, G. (10/8/2004) 'Presidents Remarks at Ask President Bush Address, Okaloo-Walton College, Niceville, Florida'. Available at: http://www.whitehouse.gov/news/releases/2004/08/20040810-15.html Accessed: 3.1.2006.

Bush, G. (11/9/2004) President's Radio Address'. Available at: http://www.whitehouse.gov/news/releases/2004/09/20040911-3.html Accessed: 3.1.2006.

Bush,G. (8/3/2005) 'President Discusses War on Terror, National Defense University, Fort Lesley J. McNair'. Available at: http://www.whitehouse.gov/news/releases/2005/03/20050308-3.html Accessed: 3.1.2006.

Bush, G. (9/7/2005) 'President's Radio Address'. Available at: http://www.whitehouse.gov/news/releases/2005/07/20050709.html Accessed: 3.1.2006.

Bush, G. (3/8/2005) 'President Discusses Second Term Accomplishments and Priorities, Gaylord Texan Resort and Convention Center, Grapevine, Texas'. Available at: http://www.whitehouse.gov/news/releases/2005/08/20050803.html Accessed: 3.1.2006.

Bush, G. (9/9/2005) 'President Remembers 9/11 Heroes at Medal of Valor Award Ceremony, The South Lawn'. Available at: http://www.whitehouse.gov/news/releases/2005/09/20050909-1.html Accessed: 15.7.2006.

Bush, G. (11/11/2005) 'President Commemorates Veterans Day, Discusses War on Terror, Tobyhanna Army Depot, Tobyhanna, Pennsylvania'. Available at:
http://www.whitehouse.gov/news/releases/2005/11/20051111-1.html Accessed: 15.7.2006.

Bush, G. (17/12/2005) 'President's Radio Address, The Roosevelt Room'. Available at: http://www.whitehouse.gov/news/releases/2005/12/20051217.html Accessed: 15.7.2006.

Bush, G. (18/12/2005) 'President's Address to the Nation, The Oval Office'. Available at: http://www.whitehouse.gov/news/releases/2005/12/20051218-2.html Accessed: 15.7.2006.

Bush, G. (22/5/2006) 'President Bush Discusses Global War on Terror, Arie Crown Theater at Lakeside Center – McCormick Place, Chicago, Illinois'. Available at: http://www.whitehouse.gov/news/releases/2006/05/20060522-1.html Accessed: 15.7.2006.

Bush, G. (16/6/2006) 'President's Radio Address'. Available at: http://www.whitehouse.gov/news/releases/2006/09/20060916.html Accessed: 18.5.2007.

Bush, G. (7/9/2006) 'President Bush Discusses Progress in the Global War on Terror, Cobb Galleria Centre, Atlanta, Georgia'. Available at: http://www.whitehouse.gov/news/releases/2006/09/20060907-2.html Accessed: 18.5.2007.

Bush, G. (23/1/2007) 'President Bush Delivers State of the Union Address, United States Capitol, Washington, DC'. Available at: http://www.whitehouse.gov/news/releases/2007/01/20070123-2.html Accessed: 18.5.2007.

Bush, G. (19/4/2007) 'President Bush Discusses the Global War on Terror in Tipp City, Ohio, Tipp City High School, Ohio'. Available at: http://www.whitehouse.gov/news/releases/2007/04/20070419-3.html Accessed: 18.5.2007.
Bush, G. (20/4/2007) 'President Bush Visits East Grand Rapids, Discusses Global War on Terror, East Grand Rapids High School, East Grand Rapids, Michigan'. Available at: http://www.whitehouse.gov/news/releases/2007/04/20070420-6.html Accessed: 19.8.2007.
Cheney, R. (16/9/2001) 'The Vice President Appears on Meet the Press with Tim Russert'. Available at: http://www.whitehouse.gov/vicepresident/news-speeches/speeches/vp20010916.html Accessed: 3.4.2006.
Cheney, R. (18/10/2001) 'Vice President Cheney Delivers Remarks at the 56th Annual Alfred E. Smith Memorial Foundation Dinner, Waldorf-Astoria Hotel, New York'. Available at: http://www.whitehouse.gov/vicepresident/news-speeches/speeches/vp20011018.html Accessed: 11.9.2004.
Cheney, R. (23/10/2001) 'The Vice President Receives the International Republican Institute's 2001 Freedom Award, The Capital Hilton, Washington, DC'. Available at: http://www.whitehouse.gov/vicepresident/news-speeches/speeches/vp20011023.html Accessed: 11.9.2001.
Cheney, R. (25/10/2001) 'Vice President Cheney Delivers Remarks to the Republican Governors Association'. Available at: http://www.whitehouse.gov/vicepresident/news-speeches/speeches/vp20011025.html. Accessed: 11.9.2004.
Cheney, R. (10/11/2001) 'Remarks by the Vice President at Celebration for the Marine Corps Birthday'. Available at: http://www.whitehouse.gov/vicepresident/news-speeches/speeches/vp20011110.html Accessed: 11.9.2004.
Cheney, R. (14/11/2001) 'Vice President Cheney on 60 Minutes II'. Available at: http://www.whitehouse.gov/vicepresident/news-speeches/speeches/vp20011114.html Accessed: 11.9.2004.
Cheney, R. (29/11/2001) 'Interview of Vice President Cheney with Diane Sawyer of ABC'. Available at: http://www.whitehouse.gov/vicepresident/news-speeches/speeches/vp20011129.html Accessed: 11.9.2004.
Cheney, R. (18/2/2002) 'Remarks by the Vice President at Miramar Marine Air Station'. Available at: http://www.whitehouse.gov/vicepresident/news-speeches/speeches/vp20020218.html Accessed: 11.9.2004.
Cheney, R. (21/2/2002) 'Remarks by the Vice President to Technology Industry Leaders, San Jose, California'. Available at: http://www.whitehouse.gov/vicepresident/news-speeches/speeches/vp20020221.html Accessed: 11.9.2004.
Cheney, R. (24/5/2002) 'Vice President's Remarks at the United States Naval Academy Commencement, United States Naval Academy, Annapolis, Maryland'. Available at: http://www.whitehouse.gov/news/releases/2002/05/20020524-25.html Accessed: 11.9.2004.
Cheney, R. (6/6/2002) 'The Vice President Delivers Remarks to the National Association of Home Builders'. Available at: http://www.whitehouse.gov/vicepresident/news-speeches/speeches/vp20020606.html Accessed: 11.9.2004.
Cheney, R. (19/7/2002) 'The Vice President Delivers Remarks at a Lunch for Texas Attorney General John Cornyn'. Available at: http://www.whitehouse.gov/vicepresident/news-speeches/speeches/vp20020719.html Accessed: 11.9.2004.
Cheney, R. (7/8/2002) 'Vice President Discusses the President's Economic Security Agenda, Fairmont Hotel, San Francisco, California'. Available at: http://

www.whitehouse.gov/news/releases/2002/08/20020807-4.html Accessed: 11.9.2004.

Cheney, R. (26/8/2002) 'Vice President Speaks at VFW 103rd National Convention, Remarks by the Vice President to the Veterans of Foreign Wars 103rd National Convention'. Available at: http://www.whitehouse.gov/news/releases/2002/08/20020826.html Accessed: 11.9.2004.

Cheney, R. (29/8/2002) 'Vice President Honors Veterans of Korean War, Marriott River Front Hotel, San Antonio, Texas'. Available at: http://www.whitehouse.gov/news/releases/2002/08/20020829-5.html Accessed: 11.9.2004.

Cheney, R. (2/10/2002) 'The Vice President Makes Remarks at the NRCC Gala Salute to Dick Armey and J. C. Watts, The Washington Hilton and Towers, Washington, DC'. Available at: http://www.whitehouse.gov/news/releases/2002/10/20021002-15.html Accessed: 11.9.2004.

Cheney, R. (24/10/2002) 'The Vice President Makes Remarks to Florida's Regional Domestic Security Task Force, Tampa, Florida'. Available at: http://www.whitehouse.gov/vicepresident/news-speeches/speeches/vp20021024.html Accessed: 11.9.2004.

Cheney, R. (9/4/2003) 'Vice President Cheney Salutes Troops, Remarks by the Vice President to the American Society of News Editors'. Available at: http://www.whitehouse.gov/news/releases/2003/04/20030409-4.html Accessed: 11.9.2004.

Cheney, R. (17/9/2003) 'Vice President's Remarks at 2003 Air Force Convention, Marriot Ballroom, Marriot Wardman Park Hotel, Washington, DC'. Available at: http://www.whitehouse.gov/news/releases/2003/09/20030917-3.html Accessed: 11.9.2004.

Cheney, R. (3/10/2003) 'Remarks by the Vice President at a Luncheon for Congressman Jim Gerlach, The Desmond Hotel and Convention Center, Malvern, Pennsylvania'. Available at: http://www.whitehouse.gov/news/releases/2003/10/20031003-9.html Accessed: 12.6.2007.

Cheney, R. (9/6/2004) 'Vice President Cheney's Remarks at State Funeral Ceremony, Remarks by the Vice President at the State Funeral of Ronald W. Reagan, Washington, DC, U.S. Capitol Rotunda'. Available at: http://www.whitehouse.gov/news/releases/2004/06/20040609-48.html Accessed: 12.6.2007.

Cheney, R. (18/12/2005) 'Vice President's Remarks at a Rally for the Troops in Iraq, Al-Asad Air Base, Al-Asad, Iraq'. Available at: http://www.whitehouse.gov/news/releases/2005/12/20051218-1.html Accessed: 11.9.2004.

Clarke, V. (28/10/2001) ASD PA Clarke Interview with CNN'. Available at: http://www.defenselink.mil/transcripts/transcript.aspx?transcriptid=2216 Accessed: 3.10.2006.

Feith, D. (21/4/2002) 'The War on Terrorism – America's War and Israel's War, Speech by Douglas J. Feith, Under Secretary of Defense for Policy, to American-Israel Public Affairs Committee'. Available at: http://www.defenselink.mil/policy/sections/public_statements/speeches/archive/former_usdp/feith/2002/april_21_02.html Accessed: 3.10.2006.

Feith, D. (13/11/2003) 'Council on Foreign Relations: Progress in the Global War on Terrorism, Washington, DC'. Available at: http://www.defenselink.mil/speeches/speech.aspx?speechid=587 Accessed: 3.10.2006.

Haass, R. (22/4/2002) 'Defining U.S. Foreign Policy in a Post-Post-Cold War World: The 2002 Arthur Ross Lecture, Remarks to the Foreign Policy Association'. Available at: http://www.state.gov/s/p/rem/9632.htm Accessed: 1.2.2007.

Kozaryn, L. (2002) 'Ridge: Terrorist Threat Persists; Nation Must Be Prepared, *American Forces Press Services*'. Available at: http://www.defenselink.mil/news/Nov2002/n11182002_200211187.html Accessed: 11.9.2004.

Migration Policy Institute (2003) *America's Challenge: Domestic Security, Civil Liberties, and National Unity after September 11*. Available at: http://www.migrationpolicy.org/pubs/Americas_Challenges.pdf Accessed: 3.1.2006.

Pentagon Memorial Project (Undated). Available at: http://memorial.pentagon.mil/ Accessed: 14.2.2006.

Powell, C. (11/9/2001) 'Corrected Version: Press Briefing on Board Plane En Route to Washington DC, Washington, DC'. Available at: http://www.state.gov/secretary/former/powell/remarks/2001/4863.htm Accessed: 11.9.2004.

Powell, C. (12/9/2001a) 'On-The-Record Briefing (1430 Hours), Washington, DC'. Available at: http://www.state.gov/secretary/former/powell/remarks/2001/4880.htm Accessed: 11.9.2004.

Powell, C. (12/9/2001b) 'Interview on Fox Morning News with Tony Snow, Washington, DC'. Available at: http://www.state.gov/secretary/former/powell/remarks/2001/4866.htm Accessed: 11.9.2004.

Powell, C. (12/9/2001c) 'Interview by National Public Radio, Washington, DC'. Available at: http://www.state.gov/secretary/former/powell/remarks/2001/4884.htm Accessed: 11.9.2004.

Powell, C. (13/9/2001) 'On-The-Record Briefing (1300 Hours), Washington, DC'. Available at: http://www.state.gov/secretary/former/powell/remarks/2001/4910.htm Accessed: 11.9.2004.

Powell, C. (14/9/2001) 'On-the-Record Press Briefing (1415 Hours), Washington, DC'. Available at: http://www.state.gov/secretary/former/powell/remarks/2001/4921.htm Accessed: 11.9.2004.

Powell, C. (16/9/2001) 'Interview on CBS' Face the Nation, Washington, DC'. Available at: http://www.state.gov/secretary/former/powell/remarks/2001/4923.htm Accessed: 11.9.2004.

Powell, C. (21/9/2001a) 'Remarks to the Special Session of the Organization of American States, Organization of American States Building, Washington, DC'. Available at: http://www.state.gov/secretary/former/powell/remarks/2001/4997.htm Accessed: 11.9.2004.

Powell, C. & Manley, J. (21/9/2001b) 'Remarks with Canadian Minister of Foreign Affairs John Manley, Washington, DC'. Available at: http://www.state.gov/secretary/former/powell/remarks/2001/5008.htm Accessed: 11.9.2004.

Powell, C. (23/9/2001) 'Interview on Meet The Press, Washington, DC'. Available at: http://www.state.gov/secretary/former/powell/remarks/2001/5012.htm Accessed: 11.9.2004.

Powell, C. (5/10/2001) 'Remarks with Participants in the International Visitors Program from Central Asia, Washington, DC'. Available at: http://www.state.gov/secretary/former/powell/remarks/2001/5322.htm Accessed: 11.9.2004.

Powell, C. (10/10/2001) 'FBI Announcement on Terrorists: Remarks at On-the-record Briefing at FBI Headquarters'. Available at: http://www.state.gov/secretary/former/powell/remarks/2001/5289.htm Accessed: 11.9.2004.

Powell, C. (19/11/2001) 'United States Position on Terrorists and Peace in the Middle East, Remarks at the McConnell Center for Political Leadership, Kentucky'. Available at: http://www.state.gov/secretary/former/powell/remarks/2001/6219.htm Accessed: 11.9.2004.

Powell, C. (29/11/2001) 'Remarks at Iftaar Dinner, Washington, DC'. Available at: http://www.state.gov/secretary/former/powell/remarks/2001/6577.htm Accessed: 11.9.2004.

Powell, C. (26/12/2001) 'Statement on Designation of Foreign Terrorist Organizations, Washington, DC'. Available at: http://www.state.gov/secretary/former/powell/remarks/2001/dec/6980.htm Accessed: 11.9.2004.

Powell, C. (22/5/2002) 'Remarks upon Arrival in Germany, Tegel International Airport, Berlin, Germany'. Available at: http://www.state.gov/secretary/former/powell/remarks/2002/10410.htm Accessed: 11.9.2004.

Powell, C. (11/7/2002) 'Homeland Security, Testimony before the House Select Committee on Homeland Security, Washington, DC'. Available at: http://www.state.gov/secretary/former/powell/remarks/2002/11793.htm Accessed: 11.9.2004.

Powell, C. (10/9/2002) 'Dedication of September 11: "Bearing Witness to History Exhibit", National Museum of American History, Smithsonian Institute, Washington, DC'. Available at: http://www.state.gov/secretary/former/powell/remarks/2002/13374.htm Accessed: 11.9.2004.

Powell, C. (7/12/2002) 'Remarks at the 25th Anniversary Kennedy Center Honors, Washington, DC'. Available at: http://www.state.gov/secretary/former/powell/remarks/2002/15785.htm Accessed: 11.9.2004.

Powell, C. (27/7/2004) 'Remarks by Secretary of State, Colin, L. Powell, and Hungarian Foreign Minister, Laszlo Kovacs at the Hungarian Chiefs of Mission Conference'. Available at: http://estonia.usembassy.gov/powellspeech.php Accessed: 3.1.2006.

Prosper, P. (20/2/2002) 'Status and Treatment of Taliban and al-Qaida Detainees, Chatham House, London'. Available at: http://www.state.gov/s/wci/us_releases/rm/2002/8491.htm. Accessed: 21.9.2006.

Rice, C. & Ridge, T. (9/10/2001) 'New Counter-Terrorism and CyberSpace Security Positions Announced'. Available at: http://www.whitehouse.gov/news/releases/2001/10/20011009-4.html Accessed: 11.9.2004.

Rice, C. (16/10/2001) 'National Security Advisor Interview with Al Jazeera TV'. Available at: http://www.whitehouse.gov/news/releases/2001/10/20011016-3.html Accessed: 9.8.2006.

Rice, C. (1/2/2002) 'National Security Advisor Speaks at Conference, Remarks by the National Security Advisor Condoleezza Rice to the Conservative Political Action Conference, Marriott Crystal Gateway, Arlington, Virginia'. Available at: http://www.whitehouse.gov/news/releases/2002/02/20020201-6.html Accessed: 11.9.2004.

Rice, C. (8/4/2002) 'National Security Advisor Speaks at Texas A&M, Remarks by Condoleezza Rice, Assistant to the President for National Security Affairs at the William Waldo Cameron Forum on Public Affairs, George Bush Presidential Library Foundation, Texas A&M University, College Station Texas'. Available at: http://www.whitehouse.gov/news/releases/2002/04/20020408-6.html Accessed: 11.9.2004.

Rice, C. (10/4/2002) 'Remarks by Condoleezza Rice, Assistant to the President for National Security Affairs, Remarks by Dr. Rice at the 2002 National Commemoration of the Days of Remembrance, U.S. Capitol Rotunda, Washington, DC'. Available at: http://www.whitehouse.gov/news/releases/2002/04/20020410-8.html Accessed: 11.9.2004.

Rice, C. (29/4/2002) 'Remarks by National Security Advisor Condoleezza Rice on Terrorism and Foreign Policy, Paul H. Nitze School of Advanced International Studies, Johns Hopkins University, Kenney Auditorium, Washington, DC'. Available at: http://www.whitehouse.gov/news/releases/2002/04/20020429-9.html Accessed: 11.9.2004.

Rice, C. (16/5/2002) 'National Security Advisor Holds Press Briefing, Press Briefing by National Security Advisor Dr. Condoleezza Rice, The James S. Brady Briefing Room'. Available at: http://www.whitehouse.gov/news/releases/2002/05/20020516-13.html Accessed: 11.9.2004.

Rice, C. (20/5/2002) 'Press Briefing by National Security Advisor Dr. Condoleezza Rice on the President's Trip to Europe and Russia, The James S. Brady Briefing Room'. Available at: http://www.whitehouse.gov/news/releases/2002/05/20020520-9.html Accessed: 11.9.2004.

Rice, C. (8/9/2002) 'CNN Late Edition with Wolf Blitzer, Interview with Condoleezza Rice; Pataki Talks About 9-11; Graham, Shelby Discuss War on Terrorism'. Available at: http://transcripts.cnn.com/TRANSCRIPTS/0209/08/le.00.html Accessed: 11.9.2004.

Rice, C. (1/10/2002) 'Dr. Condoleezza Rice Discusses President's National Security Strategy, Waldorf Astoria Hotel, New York, New York'. Available at: http://www.whitehouse.gov/news/releases/2002/10/20021001-6.html Accessed: 11.9.2004.

Rice, C. (4/12/2002) 'Remarks by National Security Advisor Condoleezza Rice at the Karamah Iftaar, Karamah, Washington, DC'. Available at: http://www.whitehouse.gov/news/releases/2002/12/20021204-17.html Accessed: 3.1.2006.

Rice, C. (25/8/2003) 'National Security Advisor Condoleezza Rice Remarks to Veterans of Foreign Wars, Remarks by Condoleezza Rice, Assistant to the President for National Security Affairs at the 104th National Convention of the Veterans of Foreign Wars, Harry B. Gonzalez Convention Center, San Antonio, Texas'. Available at: http://www.whitehouse.gov/news/releases/2003/08/20030825-1.html accessed: 11.9.2004.

Rice, C. (31/10/2003) 'National Security Advisor Dr. Rice Remarks to National Legal Center, Remarks by National Security Advisor Dr. Condoleezza Rice to the National Legal Center for the Public Interest, The Waldorf Astoria Hotel, New York, New York'. Available at: http://www.whitehouse.gov/news/releases/2003/10/20031031-5.html Accessed: 11.9.2004.

Rice, C. (20/6/2005) 'Remarks at the American University in Cairo, Cairo, Egypt'. Available at: http://www.state.gov/secretary/rm/2005/48328.htm Accessed: 3.1.2006.

Ridge, T., Meehan, P., Nolan, J. & Fauci, A. (30/10/2001) 'Tuesday's Homeland Security Briefing, Press Briefing by Director of the Office of Homeland Security, Governor Tom Ridge; Director of Emergency Environmental Health Service, CDC, Dr. Pat Meehan; Deputy Postmaster General John Nolan; National Institution of Health, Dr. Anthony Fauci, The James S. Brady Briefing Room'.

Available at: http://www.whitehouse.gov/news/releases/2001/10/20011030-3.html Accessed: 3.1.2006.

Ridge, T., Mueller, R. Nolan, J., & Cohen, M. (2/11/2001) 'Friday's Homeland Security Briefing: Press Briefing by Homeland Security Director Tom Ridge, FBI Director Bob Mueller, Deputy Postmaster General John Nolan and Director of Bacterial and Mycotic Disease for the CDC Mitch Cohen, The James S. Brady Briefing Room'. Available at http://www.whitehouse.gov/news/releases/2001/11/20011102-3.html Accessed: 3.1.2006.

Ridge, T. & Whitman, C. (9/11/2001) 'Friday's Homeland Security Briefing, Press Briefing by Director of Homeland Security Governor Tom Ridge and EPA Administrator Christine Todd Whitman, The James S. Brady Press Briefing Room'. Available at: http://www.whitehouse.gov/news/releases/2001/11/20011109-5.html Accessed: 11.9.2004.

Ridge, T. (27/11/2001) 'Governor Ridge Speaks at Homeland Security and Defense Conference'. Available at: http://www.dhs.gov/dhspublic/display?content=161 Accessed: 3.1.2006.

Ridge, T. (11/4/2002) 'Remarks by Homeland Security Director Tom Ridge to the American Society of Newspaper Editors, Washington, DC'. Available at: http://www.dhs.gov/xnews/speeches/speech_0050.shtm Accessed: 3.1.2006.

Ridge, T. (23/4/2002) 'Remarks by Homeland Security Director Tom Ridge to the Electronics Industries Alliance, Grand Hyatt Hotel, Washington, DC'. Available at: http://www.dhs.gov/xnews/speeches/speech_0049.shtm Accessed: 3.1.2006.

Ridge, T. (10/6/2002) 'Remarks by Homeland Security Director Tom Ridge to the National Association of Broadcasters Education Foundation 2002 Service to America Summit, Ronald Reagan Building, Washington, DC'. Available at: http://www.dhs.gov/xnews/speeches/speech_0045.shtm Accessed: 3.1.2006.

Ridge, T. (26/9/2002) 'Director Ridge Address U.S. Conference of Mayors, Remarks by Homeland Security Advisor Tom Ridge on the Department of Homeland Security, Indian Treaty Room, Dwight D. Eisenhower Executive Office Building'. Available at: http://www.dhs.gov/xnews/speeches/speech_0025.shtm Accessed: 3.1.2006.

Rumsfeld, D. Shelton, H., White, T., Levin, C. & Warner, J. (11/9/2001) 'News Briefing on Pentagon Attack, The Pentagon'. Available at: http://www.state.gov/coalition/cr/rm/2001/5002.htm Accessed: 11.9.2004.

Rumsfeld, D. Shelton, H., Clarke, V. & Abell, C. (27/9/2001) 'DoD News Briefing – Secretary Rumsfeld'. Available at: http://www.defenselink.mil/transcripts/transcript.aspx?transcriptid=1939 Accessed: 11/9/2004.

Rumsfeld, D. (11/10/2001) 'Memorial Service in Remembrance of Those Lost on September 11th'. Available at: http://www.defenselink.mil/Speeches/Speech.aspx?SpeechID=448 Accessed: 11/9/2003.

Rumsfeld, D. (1/11/2001) 'Statement of the Secretary of Defense'. Available at: http://www.defenselink.mil/releases/release.aspx?releaseid=3142 Accessed: 11.9.2004.

Rumsfeld, D. (15/11/2001) 'Thanksgiving Message to the Troops from the Secretary of Defense'. Available at: http://www.defenselink.mil/news/Nov2001/n11152001_200111154.html Accessed: 11.9.2004.

Rumsfeld, D. & Myers, R. (11/12/2001) 'Ceremony for Remembrance, The Pentagon, Washington, DC'. Available at: http://www.defenselink.mil/speeches/speech.aspx?speechid=503 Accessed: 11/9/2004.

Rumsfeld, D. & Wolfowitz, P. (12/12/2001) 'Prepared Statement: Senate Armed Services Committee "Military Commissions"'. Available at: http://www.defenselink.mil/speeches/speech.aspx?speechid=506 Accessed: 11.9.2004.

Rumsfeld, D. (18/12/2001) 'Secretary Rumsfeld News Briefing in Brussels'. Available at: http://www.yale.edu/lawweb/avalon/sept_11/rumsfeld_007.htm Accessed: 11.9.2004.

Rumsfeld, D. (31/1/2002) 'Secretary Rumsfeld Speaks on "21st Century Transformation" of US Armed Forces (Transcript of Remarks and Question and Answer Period), as Delivered by Secretary of Defence Donald Rumsfeld, National Defense University, Fort McNair, Washington, DC'. Available at: http://www.defenselink.mil/speeches/2002/s20020131-secdef.html Accessed: 3.10.2006.

Rumsfeld, D. (4/2/2002) 'Secretary Rumsfeld Interview with Jim Lehrer, PBS "Newsmaker"'. Available at: http://www.defenselink.mil/transcripts/transcript.aspx?transcriptid=2606 Accessed: 3.10.2006.

Rumsfeld, D. & Myers, R. (12/2/2002) 'DoD News Briefing, Secretary Rumsfeld and Gen. Myers'. Available at: http://www.defenselink.mil/transcripts/transcript.aspx?transcriptid=2636 Accessed: 11.9.2004.

Rumsfeld, D. (6/6/2002) 'Secretary Rumsfeld Press Conference at NATO Headquarters, Brussels, Belgium'. Available at: http://www.defenselink.mil/transcripts/transcript.aspx?transcriptid=3490 Accessed: 11.9.2004.

Rumsfeld, D. (9/6/2002) 'Presenter: Secretary of Defense Donald H. Rumsfeld, Camp Doha Town Hall Meeting, Kuwait City, Kuwait'. Available at: http://www.defenselink.mil/transcripts/2002/t06102002_t0609thm.html Accessed: 9.1.2007.

Rumsfeld, D. (6/8/2002) 'Pentagon Town Hall Meeting, Remarks as Delivered by Secretary of Defense Donald H. Rumsfeld, Pentagon, Washington, DC'. Available at: http://www.defenselink.mil/Speeches/Speech.aspx?SpeechID=274 Accessed: 11.9.2004.

Rumsfeld, D. (11/9/2002) 'Secretary Rumsfeld Interview with Associated Press'. Available at: http://www.defenselink.mil/transcripts/transcript.aspx?transcriptid=3643 Accessed: 11.9.2004.

Rumsfeld, D. (12/9/2002) 'Arlington National Cemetery Funeral Service for the Unidentified Victims of the Attack on the Pentagon, Remarks by Secretary of Defense Donald H. Rumsfeld, Arlington National Cemetery'. Available at: http://www.defenselink.mil/speeches/speech.aspx?speechid=281 Accessed: 11.9.2004.

Rumsfeld, D. (7/9/2004) 'Secretary Rumsfeld's Interview with WHO-AM Radio's Jan Mikelson'. Available at: http://www.defenselink.mil/transcripts/transcript.aspx?transcriptid=2354 Accessed: 5.6.2006.

Rumsfeld, D. (9/9/2004) 'Secretary Rumsfeld's Patriot Day Radio Message to the Troops'. Available at: http://www.defenselink.mil/news/newsarticle.aspx?id=25367 Accessed: 5.6.2006.

Rumsfeld, D. (11/9/2004) 'Secretary of Defense Donald Rumsfeld Remarks at Wreath Laying Ceremony at Arlington National Cemetery'. Available at:

http://www.defenselink.mil/releases/release.aspx?releaseid=7725 Accessed: 5.6.2006.

Rumsfeld, D. (2/8/2005) 'Secretary Rumsfeld Remarks at Greater Dallas Chamber of Commerce'. Available at: http://www.defenselink.mil/transcripts/transcript.aspx?transcriptid=3856 Accessed: 5.6.2006.

Rumsfeld, D. (27/3/2006) 'Army War College, as Delivered by Secretary of Defense Donald H. Rumsfeld, Carlisle Barracks, Pennsylvania'. Available at: http://www.defenselink.mil/speeches/speech.aspx?speechid=22 Accessed: 5.6.2006.

Taylor, F. (25/9/2001) 'Terrorist Threats against America, Testimony to the Committee on International Relations, Washington DC' Available at: http://www.state.gov/s/ct/rls/rm/2001/5215.htm Accessed: 14.7.2006.

Taylor, F. (10/10/2001) 'The Presence of International Terrorist Groups in the Western Hemisphere: Remarks before the Committee on International Relations, Subcommittee on the Western Hemisphere Committee, US House of Representative'. Available at: http://www.state.gov/s/ct/rls/rm/2001/5674.htm Accessed: 5.4.2006.

Taylor, F. (23/10/2001). 'The Importance of International Counterterrorism Treaties, Testimony before the Committee on Foreign Relations, United States Senate, Washington, DC'. Available at: http://www.state.gov/s/ct/rls/rm/2001/5662.htm Accessed: 11.9.2004.

Taylor, F. (26/10/2001) 'A Discussion on the Global Campaign against Terrorism, Remarks before the National Foreign Policy Conference for Leaders of Nongovernmental Organizations, Washington, DC'. Available at: http://www.state.gov/s/ct/rls/rm/2001/5773.htm Accessed: 11.9.2004.

The White House (9/2002) *The National Security Strategy of the United States of America*. Available at: http://www.whitehouse.gov/nsc/nss/2002/nss.pdf Accessed: 11.9.2004.

The White House (2/2003) *The National Strategy for Combating Terrorism*. Available at: http://www.whitehouse.gov/news/releases/2003/02/counter_terrorism/counter_terrorism_strategy.pdf Accessed: 11.9.2004.

The White House (17/3/2003) 'Fact Sheet: Operation Liberty Shield'. Available at: http://www.whitehouse.gov/news/releases/2003/03/20030317-9.html Accessed: 11.9.2004.

The White House (29/4/2003) 'Operation Enduring Freedom: One Year of Accomplishments'. Available at: http://www.whitehouse.gov/infocus/defense/enduringfreedom.html Accessed: 11.9.2004.

The White House (6/11/2003) 'President Bush Calls for a "Forward Strategy of Freedom" to Promote Democracy in the Middle East: Today's Presidential Action'. Available at: http://www.whitehouse.gov/news/releases/2003/11/20031106-11.html Accessed: 20.4.2007.

The White House (9/2006) *9/11 Five Years Later: Successes and Challenges*. Available at: http://www.whitehouse.gov/nsc/waronterror/2006/waronterror0906.pdf Accessed: 20.4.2007.

Thompson, L. (2/11/2001) 'Final Remarks of Deputy Attorney General Larry Thompson to the ABA White Collar Crime Committee'. Available at: http://www.usdoj.gov/archive/dag/speeches/2001/110201abawccspeechfinal.htm Accessed: 11.9.2004.

Thompson, L. (8/11/2001) 'Final Remarks of Deputy Attorney General Larry Thompson to ALI-ABA Course on Criminal Enforcement of Environmental Laws. What Can Be Expected from the New Administration?, Westin Fairfax, Washington, DC'. Available at: http://www.usdoj.gov/archive/dag/speeches/2001/110801aliabaenvcrimesspeech.htm Accessed: 11.9.2004.

Thompson, L. (12/3/2002) 'Final Remarks of Deputy Attorney General Larry Thompson, INS Commissioner's Conference, Argent Hotel, San Francisco, California'. Available at: http://www.usdoj.gov/archive/dag/speeches/2002/031202insspeech.htm Accessed: 11.9.2004.

Thompson, L. (15/3/2002) 'Final Remarks of Deputy Attorney General Larry Thompson at the American College of Trial Lawyers'. Available at: http://www.usdoj.gov/archive/dag/speeches/2002/031502actlspeech.htm Accessed: 11.9.2004.

Thompson, L. (2/5/2002) 'Final Speech of Deputy Attorney General Larry Thompson at the Economic Club of Flordia, Tallahassee, Florida'. Available at: http://www.usdoj.gov/archive/dag/speeches/2002/050202economicclubofflorida.htm Accessed: 11.9.2004.

Wolfowitz, P. (3–4/10/2001) 'Prepared Statement for the House and Senate Armed Services Committees: "Building a Military for the 21st Century"'. Available at: http://www.defenselink.mil/speeches/2001/s20011003-depsecdef.html Accessed: 4.4.2006.

Wolfowitz, P. (14/11/2001) 'Fletcher Conference, Remarks by Deputy Secretary of Defense Paul Wolfowitz, Ronald Raegan Building and International Trade Center'. Available at: http://www.defenselink.mil/speeches/speech.aspx?speechid=497 Accessed: 11.9.2004.

Wolfowitz, P. (31/1/2002) 'Remarks by Deputy Secretary of Defense Paul Wolfowitz, The Pentagon, Washington, DC'. Available at: http://www.defenselink.mil/speeches/2002/s20020131-depsecdef.html Accessed: 3.10.2006.

Wolfowitz, P. (31/5/2002) 'Deputy Secretary Wolfowitz Luncheon Press Event in Singapore' Available at: http://www.defenselink.mil/transcripts/transcript.aspx?transcriptid=3472 Accessed: 11.9.2004.

Wolfowitz, P. (1/6/2002) '"The Gathering Storm: The Threat of Global Terror and Asia/Pacific Security", as Prepared for Delivery by US Deputy Secretary of Defense Paul Wolfowitz, Singapore'. Available at: http://www.defenselink.mil/speeches/speech.aspx?speechid=248 Accessed: 11.9.2004.

Wolfowitz, P. (19/9/2002) 'Joint Inquiry Hearing on Counterterrorist Center Customer Perspective, Prepared Testimony of Deputy Secretary of Defense Paul Wolfowitz'. Available at: http://www.defenselink.mil/speeches/speech.aspx?speechid=285 Accessed: 11.9.2004.

Wolfowitz, P. (2/12/2002) 'International Institute for Strategic Studies, Remarks by Deputy Secretary of Defense Paul Wolfowitz, Arundel House, London'. Available at: http://www.defenselink.mil/speeches/speech.aspx?speechid=309 Accessed: 11.9.2004.

Index